The Gamma

A Study of Cell-Organelle Interactions in the Development of the Water Mold *Blastocladiella emersonii*

R. B. MYERS and E. C. CANTINO

Department of Botany and Plant Pathology,
Michigan State University, East Lansing, Mich.

43 figures and 12 tables

S. Karger · Basel · München · Paris · London · New York · Sydney

Monographs in Developmental Biology

Vol. 3
CHEN, P.S. (Zürich): Biochemical Aspects of Insect Development.
VIII + 236 p., 25 fig., 8 tab., 1971. ISBN 3-8055-1265-1

Vol. 4
CARLSON, B.M. (Ann Arbor, Mich.): The Regeneration of Minced Muscles.
VII + 128 p., 49 fig., 2 tab., 1972. ISBN 3-8055-1410-7

Vol. 5
ZOTIN, A.I. (Moscow): Thermodynamic Aspects of Developmental Biology.
X + 159 p., 30 fig., 23 tab., 1972. ISBN 3-8055-1411-5

Vol. 6
MILMAN, L.S. and YUROWITZKY, Y.G. (Moscow): Regulation of Glycolsyis in the Early Development of Fish Embryos.
VI + 106 p., 23 fig., 25 tab., 1973. ISBN 3-8055-1469-7

Vol. 7
POGLAZOV, B.F. (Moscow): Morphogenesis of T-Even Bacteriophages.
VI + 105 p., 56 fig., 6 tab., 1973. ISBN 3-8055-1645-2

S. Karger · Basel · München · Paris · London · New York · Sydney
Arnold-Böcklin-Strasse 25, CH-4011 Basel (Switzerland)

Printed in Switzerland by Gasser & Eggerling AG, Chur
ISBN 3-8055-1735-1

Contents

Dedication and Acknowledgement

We dedicate this monograph firstly to RALPH EMERSON, whose rare and contagious enthusiasm kindled in his first disciple the curious urges that led to the discovery of his namesake, and secondly to our many colleagues who, motivated by *B. emersonii's* attributes and succumbing to its charm, have helped to chart many interlocking facets of its personality.

We are grateful to GARY MILLS for his kind and invaluable help during our labors. Most of the unpublished work done in our laboratory and reported herein was sustained by various general research support grants to E. C. C. from the National Institutes of Health and the National Science Foundation.

I. Introduction

The water mold *Blastocladiella emersonii* was discovered in a freshwater pond on the University of Pennsylvania campus and isolated in pure culture in 1949 [CANTINO, 1951]; different aspects of the studies done with it since that time have been reviewed comprehensively by us [CANTINO and LOVETT, 1964; CANTINO, 1966; CANTINO *et al.*, 1968; TRUESDELL and CANTINO, 1971] and by others [BALDWIN and RUSCH, 1965; TURIAN, 1969; SMITH and GALBRAITH, 1971]. The only known method of reproduction occurs by way of motile, uniflagellate, and generally uninucleate swarm-cells, i.e. zoospores.[1] For some two decades, it has been known [CANTINO and HYATT, 1953a; CANTINO, 1966] that zoospores produced by the wild type strain of *B. emersonii* give rise to coenocytic vegetative thalli which then differentiate into several phenotypes distinguishable from one another by means of various criteria (fig. 1). Ordinarily, such a population will consist of a large class (about 98–99% of the plants) of relatively rapidly growing thalli, each of which produces a terminal ordinary colorless (OC) sporangium (hence, OC plant, OC pathway). However, on certain media, as OC plants are maturing, a small class (about 1–2% of the population) of thalli with slightly longer generation times appears among them; they produce either orange (O) or late colorless (LC) sporangia (hence, O and LC plants, O and LC pathways) in roughly equal numbers. Orange plants are generally larger than OC plants; LC plants are generally smaller. Finally, if germlings are produced under suitable starvation conditions [HENNESSY and CANTINO, 1972], they give rise (fig. 1, top) to 'miniplants' that produce but a single zoospore

[1] A reasonable case can be built [CANTINO, 1966; HORENSTEIN and CANTINO, 1969; MATSUMAE and CANTINO, 1970] to support the notion that some swarm-cells, insofar as they are produced by male-like and female-like thalli (i.e. the O and LC plants discussed in this paragraph) may be male and female sex cells, respectively, albeit nonfunctional ones in that they do not indulge in conventional gametic copulation. However, to simplify the various presentations in this monograph, all swarm-cells formed by *B. emersonii*, irrespective of their immediate parental phenotypes, will simply be called 'zoospores' – or just 'spores'.

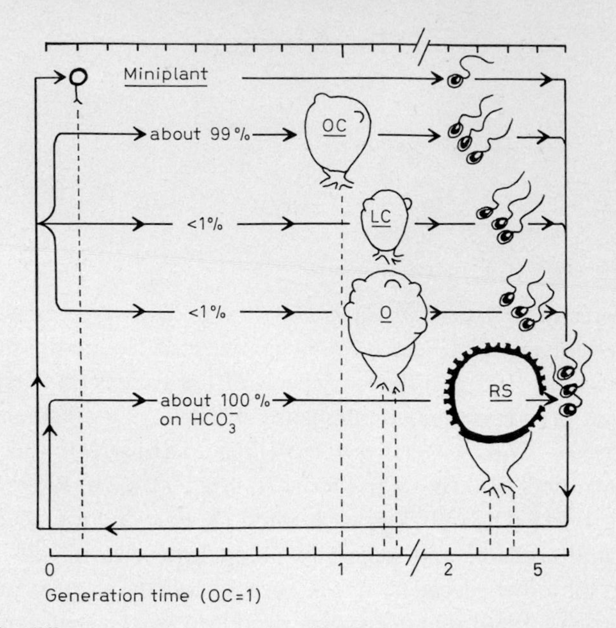

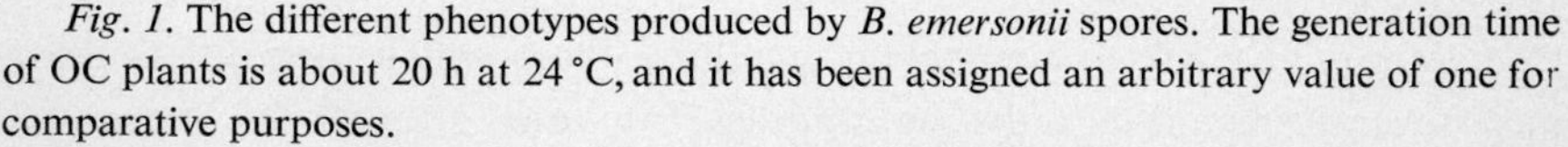

Fig. 1. The different phenotypes produced by *B. emersonii* spores. The generation time of OC plants is about 20 h at 24 °C, and it has been assigned an arbitrary value of one for comparative purposes.

(hence, microplant, microcycle). Alternatively, if plants are grown on bicarbonate media, they give rise to a fifth phenotype bearing a brown, pitted, thick-walled, resistant sporangium (RS; hence, RS plant, RS pathway) with a much longer generation time.

When individual, randomly selected OC, O, or LC plants are isolated, and the zoospores produced by them are put on fresh media, the ensuing generations often contain all three phenotypes again (fig. 1, arrows). In any single population, the relative proportions of the various phenotypes produced depend, in part, on the nature of the parent plant, i.e., whether it is an LC, OC, O, or RS plant [Cantino and Hyatt, 1953a]; proportions can also be affected by its age [Cantino, 1969] and its previous history [Cantino and Hyatt, 1953a]. But in spite of such phenotypic variability in wild type *B. emersonii*, no conventional, i.e. irreversible, fusions have ever been detected [Cantino and Horenstein, 1954] among its motile cells. Consequently, sex does not appear to be responsible for the variability. Since these zoospores are almost always uninucleate, the obvious question from the start has been: by what mechanism does such variability arise in a population of apparently identical zoospores?

Early efforts to interpret these and related observations led us to hypothesize [CANTINO and HYATT, 1953a] that a cytoplasmic factor of unknown nature might have been causally involved: that low concentrations of the factor induced the genesis of O thalli; intermediate concentrations, OC thalli; and high concentrations, LC thalli. We labeled this hypothetical factor 'gamma' because it had been established [CANTINO and HYATT, 1953b] that the orange color of an O plant was probably due to the presence of γ-carotene.

II. Discovery of the Gamma Particle

Subsequent examinations of zoospores stained with the Nadi reagent (dimethyl-p-phenylenediamine and α-naphthol) revealed the presence (fig. 2) of deeply stained, cytoplasmic granules about 0.5 μm in diameter [CANTINO and HORENSTEIN, 1954, 1956]. It was then proposed that these visible, Nadi-positive, cytoplasmic granules might correspond to the hypothetical cytoplasmic factor, gamma. The visible granules were therefore named 'gamma particles'.[2]

Spores from O, OC, and LC plants were then scored for their content of gamma particles [CANTINO and HORENSTEIN, 1956]. The frequency distribution curves and mean values for the number of gamma particles per spore were different for all three phenotypes (fig. 4, left). For comparative purposes, some recent tabulations using a different staining procedure [MATSUMAE *et al.*, 1970], (fig. 3, legend) are also included (fig. 4, right). It is an interesting fact that after some two decades of continuous cultivation, *B. emersonii* still displays these distinct distribution patterns, with means of approximately 8, 12, and 16 gamma particles per spore, associated with the three parental phenotypes.

Finally, the results of two other types of experiments which lent further support to our notions about the function of the gamma particles deserve to be mentioned in this introduction to this cytoplasmic organelle (fig. 5). One of these involved the effect of cycloheximide (fig. 5, left). When populations of spores from OC plants carrying the usual average of about 12 gamma particles were put on appropriate media containing about 0.15 μg cycloheximide/ml [CANTINO and HYATT, 1953a], large numbers of O plants – 25–50% rather than $<1\%$ of the population – were formed. The spores derived from these cycloheximide-induced O plants now carried the reduced number of gamma particles expected for naturally occurring O plants, i.e. about eight.

[2] Photographs were not taken at the time that Nadi-positive particles were being studied in the 1950s; we illustrate them here (fig. 3) with a recent representative picture of a zoospore derived from a variant strain of *B. emersonii* that produces an exceptionally large number of these particles.

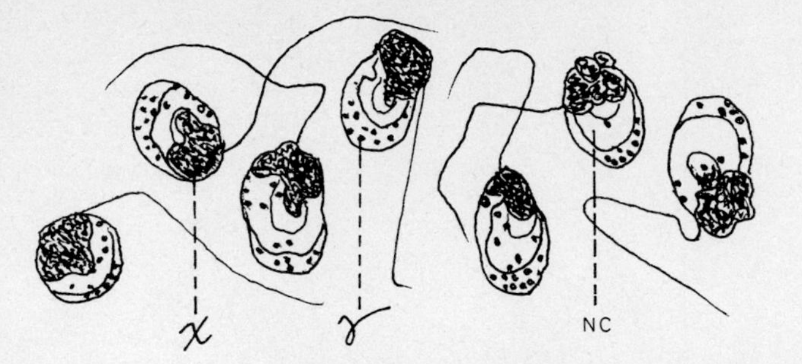

Fig. 2. The location and appearance of gamma particles as first seen in the form of Nadi-positive cytoplasmic granules (γ) in zoospores of *B. emersonii*; facsimile of representative sketches made in 1955 (Cantino and Horenstein, University of Pennsylvania). NC = Nuclear cap; χ = an aggregate containing lipoidal material and labeled the 'side body' before its components had been identified by electron microscopy (see chapter III).

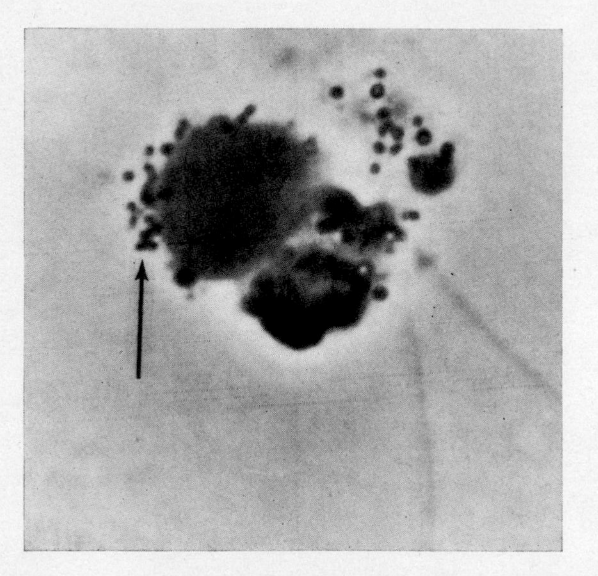

Fig. 3. Biflagellate zoospore of a mitomycin-induced orange variant (MC-3M) of *B. emersonii* [Matsumae and Cantino, 1970]. This spore, which contains two nuclei, two nuclear caps, and an abnormally large number (about 40–50) of gamma particles, can be used advantageously to illustrate the general apperance of these particles. Zoospores were chilled, fixed with 2% OsO_4, and stained with 0.25% alcoholic methylene blue; under these conditions, the expanded plasma membrane is often destroyed (as in this instance), but the fixed gamma particles remain behind sufficiently dispersed so that they can be easily counted. Gamma particles stained with methylene blue and the Nadi reagent are indistinguishable from one another in the light microscope.

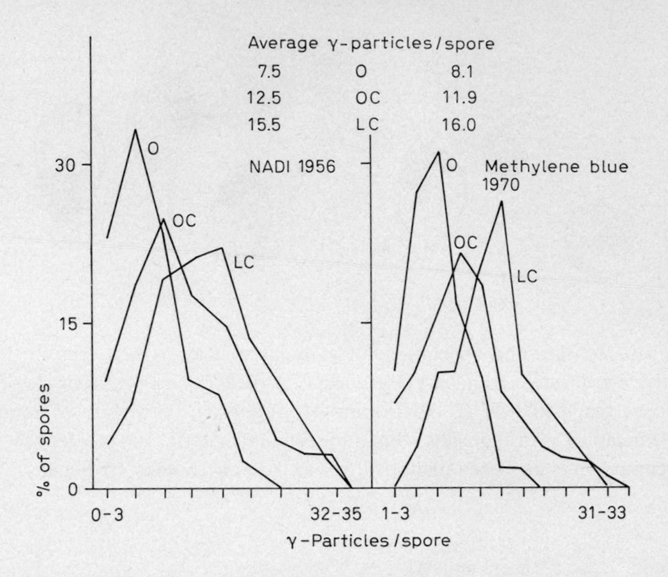

Fig. 4. The number of gamma particles in spores derived from O, OC, and LC pheno types of *B. emersonii* (see text for explanation).

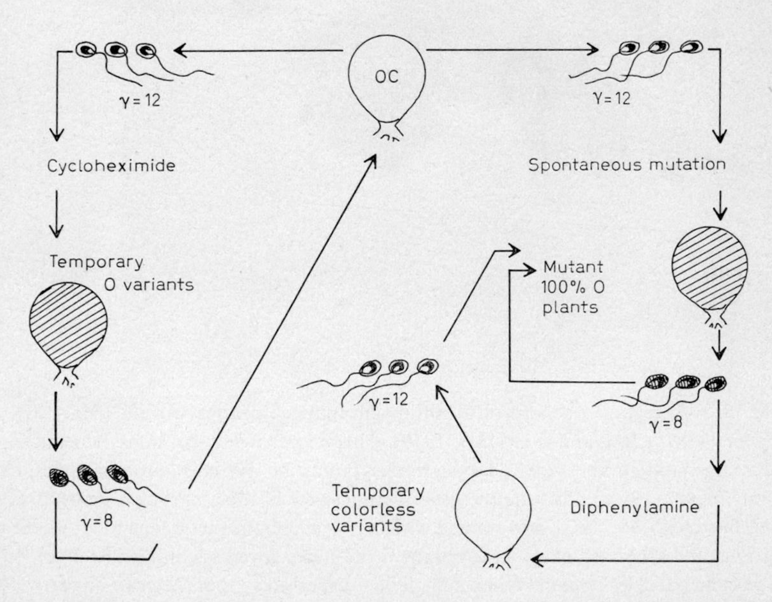

Fig. 5. Effects of cycloheximide and diphenylamine on the number of gamma particles pei spore (see text for explanation).

Many of these induced orange spores were not viable. However, when such spore populations were put back on media lacking cycloheximide, the next generation of plants consisted once again of the usual proportion of OC thalli, i.e. about 98–99%, and the spores derived from these OC plants carried about 12 gamma particles.

The other type of experiment (fig. 5, right) involved the use of a spontaneous mutant [CANTINO and HYATT, 1953b] in which *all* plants that reached maturity were orange due to the presence of γ-carotene, i.e. they were all O phenotypes. In this case, the orange spores released from them had an average of about 8 gamma particles, as had been the case for the $<1\%$ O plants (fig. 1) produced by the wild type. This naturally occurring orange mutant (unlike the temporary, cycloheximide-induced orange variants) was stable, and each successive generation of spores also carried an average of about 8 gamma particles. However, when such spores were put on media containing suitable concentrations of diphenylamine (fig. 5, right), almost all the plants in the next generation were colorless, i.e. they resembled OC phenotypes, and the spores derived from them now carried an average of about 12 gamma particles. Yet, when similar spores were put back on media lacking diphenylamine, O plants were regenerated which released spores with an average of about 8 gamma particles. Once again, therefore, it was evident that the number of gamma particles per spore was related to the nature of the parental phenotype, but it was also clear that this relationship could be modulated by certain environmental conditions imposed upon the fungus during its growth.

III. Structure of the Gamma Particle

A more exhaustive characterization of the gamma particle had to await electron-microscopic studies of the zoospore [CANTINO *et al.*, 1963]. Figure 6 illustrates the principal facts now known about the spore's internal morphology, while some of its fine structural features are seen in figure 7; further details are provided in the legends to both figures. One aspect of the spore's organization that is especially difficult to visualize from two-dimensional representations concerns the structural interrelationships that occur among the mitochondrion, lipid globules, and SB matrix; a three-dimensional view of the postition taken by these organelles, as we have interpreted it from serial sections, is presented in figure 8. For reviews and references about form and structure of the zoospore of *B. emersonii*, the reader is referred to articles by CANTINO *et al.* [1968], CANTINO and TRUESDELL [1970], and TRUESDELL and CANTINO [1972].

The cytoplasm of the zoospore contains only one kind of prominent organelle about 0.5 μm in diameter (fig. 7a, solid arrow). At the time they were discovered [CANTINO *et al.*, 1963], it seemed a virtual certainty that these bodies had to be the gamma particles seen 8 years earlier (fig. 2) with the light microscope. When fixed with OsO_4 and/or plumbic, uranyl, and permanganate salts, they became intensely stained [CANTINO and MACK, 1969]. The most common and distinguishing feature of a gamma particle profile is the horseshoe-like shape (fig. 7) of its matrix (gamma matrix). Its appearance, however, depends in part upon the fixative used. With $KMnO_4$ fixation (fig. 9a), the gamma particle's surrounding unit membrane (GS membrane) always lies close to its deeply stained matrix, and a region of the GS membrane invaginates deeply into it at the open end of the horseshoe. With glutaraldehyde fixation, however (fig. 9b), the GS membrane is always well separated from the gamma matrix. With either fixative, the matrix is usually differentiated into two regions of visibly different density. For preservation of the GS membrane (fig. 9c, solid arrow), a solution of Ur acetate and OsO_4 is the fixative of choice.

On the basis of the appearance of serial sections through the various axes

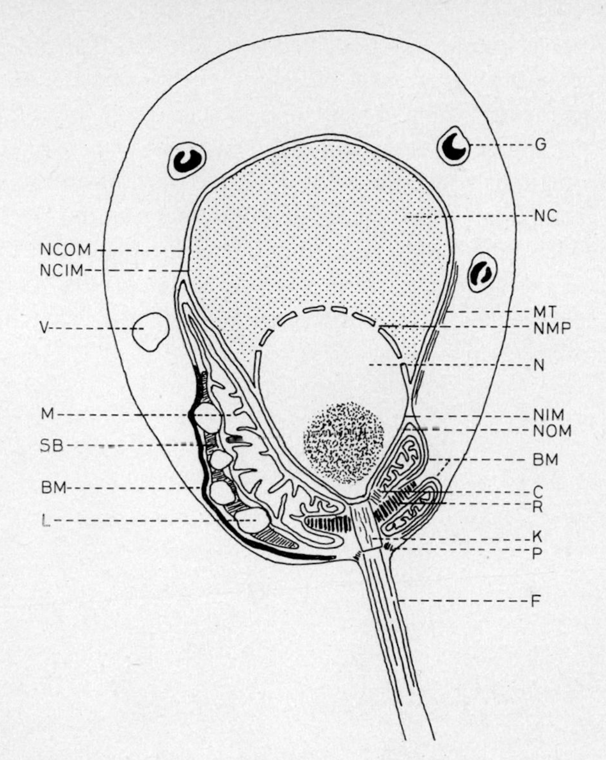

Fig. 6. Guide to the internal structure of the zoospore of *B. emersonii*. NCOM = Nuclear cap outer membrane; NCIM = nuclear cap inner membrane; V = vacuole; M = mitochondrion; SB = SB matrix of lipid sac; BM = backing membrane; L = lipid globule; G = gamma particle; NC = nuclear cap; MT = member of microtubule triplet; NMP = nuclear membrane pore; N = nucleus; NIM = nuclear inner membrane; NOM = nuclear outer membrane; C = centriole; R = rootlet; K = kinetosome; F = flagellum; P = 'prop' connecting K to plasma membrane of spore. (Note: sketch is a *composite* representation.)

of a gamma particle, a model (fig. 10) was constructed depicting its apparent three-dimensional structure. The gamma particle consists of two major components. One is an inner matrix, intensely electron-opaque and shaped like a short, open-ended canoe or hemiellipsoid, but with two additional holes at its opposite ends. Two views of the matrix are shown here (second tier from top, two sketches at left), one at right angles to its long axis, the other at right angles to its short axis. Around it (top tier, two sketches at left), is the other component, the continuous GS membrane which completely envelopes the gamma matrix. At top right (Mn), these two components – the cup-shaped matrix and the outer unit membrane – are combined to yield a reconstructed gamma particle as it appears when fixed with $KMnO_4$. The outer membrane surrounds the matrix tightly, and those parts of it opposite the holes in the cup invaginate through them to yield an organelle displaying a large concavity and two smaller ones. Yet the gamma particle can also be reconstructed to reflect other methods of fixation. For example, turning to the other sketch (Glut), if a gamma particle is fixed with glutaraldehyde, the GS membrane is distended and lies more or less free from the matrix; hence, the invaginations disappear and the gamma particle takes on an ellipsoidal shape. Sections through it yield profiles, as idealized at the bottom of the sketch, resembling authentic profiles (as seen in figure 9).

Fig. 7. a Overall view of a zoospore of *B. emersonii*; a median longitudinal section showing the nuclear cap, nucleus, nucleolus, gamma particles (solid arrows), and the single eccentrically placed mitochondrion (dotted arrow) with its adjacent lipid sac (lipid globules, SB matrix, and backing membrane). *b* Cross section through posterior end of a spore and through the vertical (i.e. relative to the long axis of the spore) mitochondrial canal housing the kinetosome and the adjacent centriole; the horizontal mitochondrial canal housing the banded rootlet is also evident. *c* An oblique longitudinal section illustrating the triplets of microtubules (arrows), pictured by LESSIE and LOVETT [1968], that extend from the region of the kinetosome upward around the nucleus (although not shown here, they extend upward around the nuclear cap as well); FULLER and CALHOUN [1968] have said that all motile cells of the *Blastocladiales* have 27 microtubules arranged in 9 groups of 3 tubules each, but they did not include illustrations of them in *B. emersonii*; our picture corroborates part of their observation, but we do not have evidence yet for the exact number of triplets present. Also visible are the kinetosome with the centriole at its upper left, part of the banded rootlet at its right, and two segments of the single mitochondrion. All spores were prepared with fixation procedure 2 of TRUESDELL and CANTINO [1970]. We are indebted to our former collaborator, Dr. L. C. TRUESDELL, who took the pictures shown here and in figures 11 and 13–15 while he was still at work in our laboratory helping to disentangle the evanescent by-products of unravelling gamma particles. Marker in $a = 1\ \mu m$; markers in b and $c = 0.5\ \mu m$.

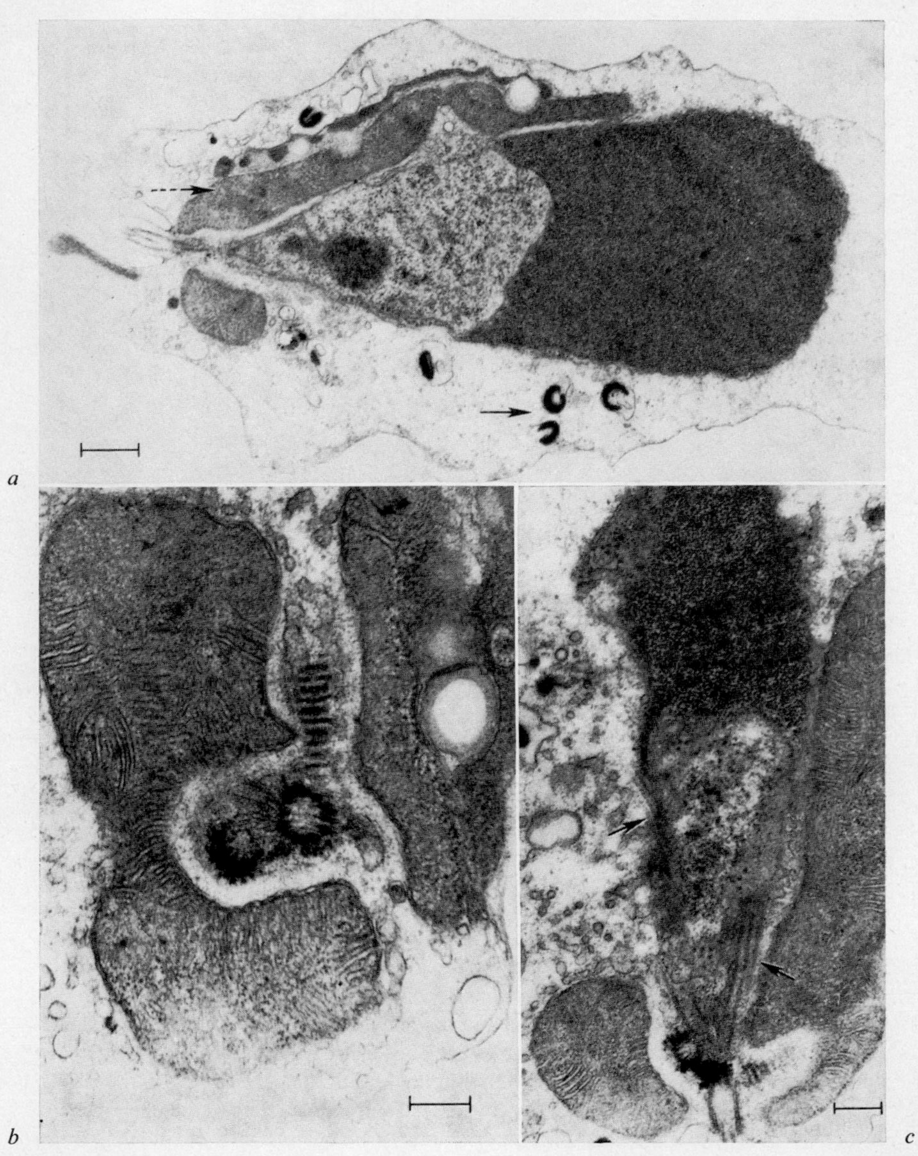

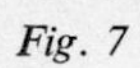

Fig. 7

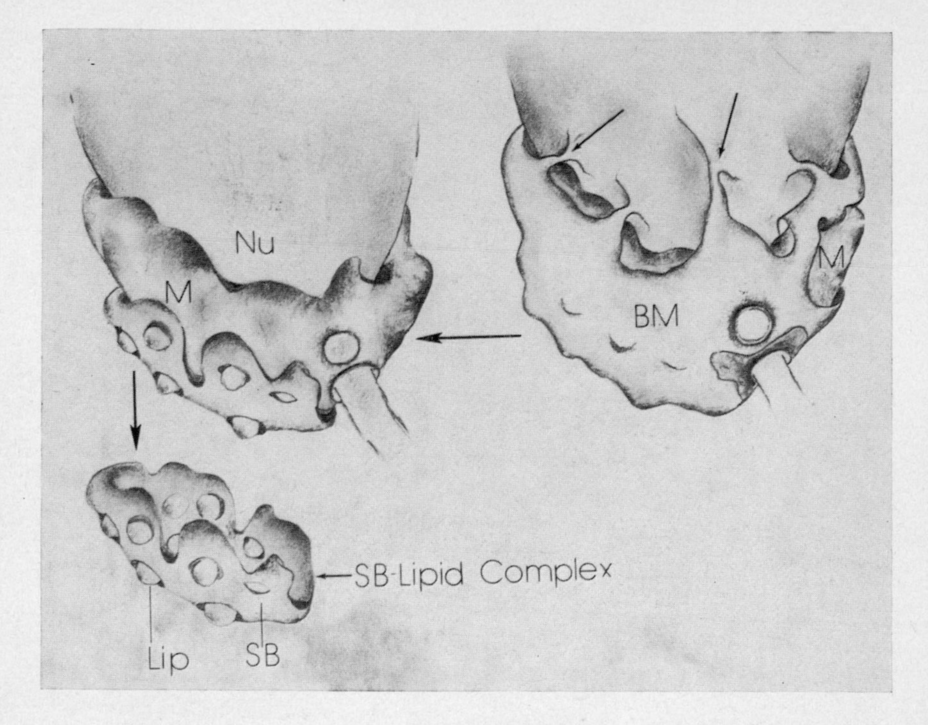

Fig. 8. The form and structural interrelationships among the main components of the 'side body' in the zoospores of *B. emersonii*, as reconstructed from serial and random sections. This tightly organized assemblage consists of a single mitochondrion (M) and a 'lipid sac'; the latter is composed of an SB matrix (SB), lipid globules (Lip), and a backing membrane (BM). Lower left: the SB-lipid complex, consisting of a unit membrane-bound SB matrix and lipid globules partially embedded therein. Upper left: the association between the SB-lipid complex and the mitochondrion, the combination lying alongside the surface of the membrane that evelopes both the nucleus (Nu) and the nuclear cap; the hole above the entry point of the flagellum is the opening to the rootlet canal in the mitochondrion. Upper right: the manner in which the mitochondrion and the SB-lipid complex are held in place and partially enclosed by the backing membrane; the latter is continuous with the outer membrane of the nucleus and nuclear cap at irregularly scattered points (arrows). At some places, the backing membrane extends downward and 'out of sight' into a cytoplasmic channel between the mitochondrion and the nuclear outer membrane; it also extends into the mitochondrial rootlet canal, it folds back under itself, and it simply terminates as a sealed edge, thus exposing the mitochondrion to the cytoplasm, as seen here in the vicinity of the mitochondrial canal housing the kinetosome and immediately to the upper right of this region. The backing membrane appears bumpy where it is tightly positioned against the lipid globules.

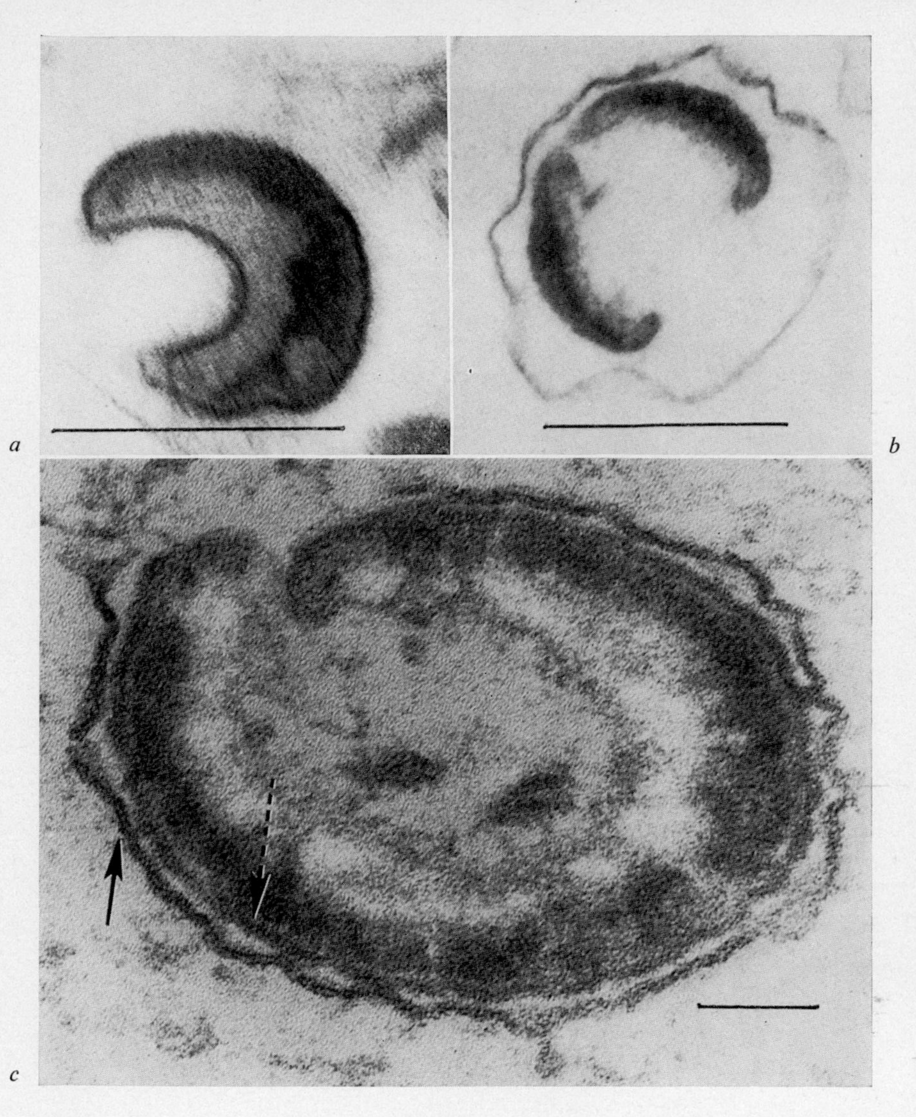

Fig. 9. The appearance of gamma particles prepared with different fixatives. 2% $KMnO_4$ (a) and 2% glutaraldehyde (b), both in 9 mM Na veronal-Na acetate at pH 7.8. *c* 0.5% glutaraldehyde and 0.25% OsO_4 (fixative 2 of TRUESDELL and CANTINO, (1970). The unit membrane around the gamma particle is distinctly tri-laminar (solid arrow) in appearance, while the outer edge (dotted arrow) of the gamma particle matrix, although usually differentiated into a more or less distinct 'line', only occasionally bears segments resembling a tri-laminar arrangement. Markers in *a* and *b* = 0.5 µm; marker in *c* = 0.1 µm).

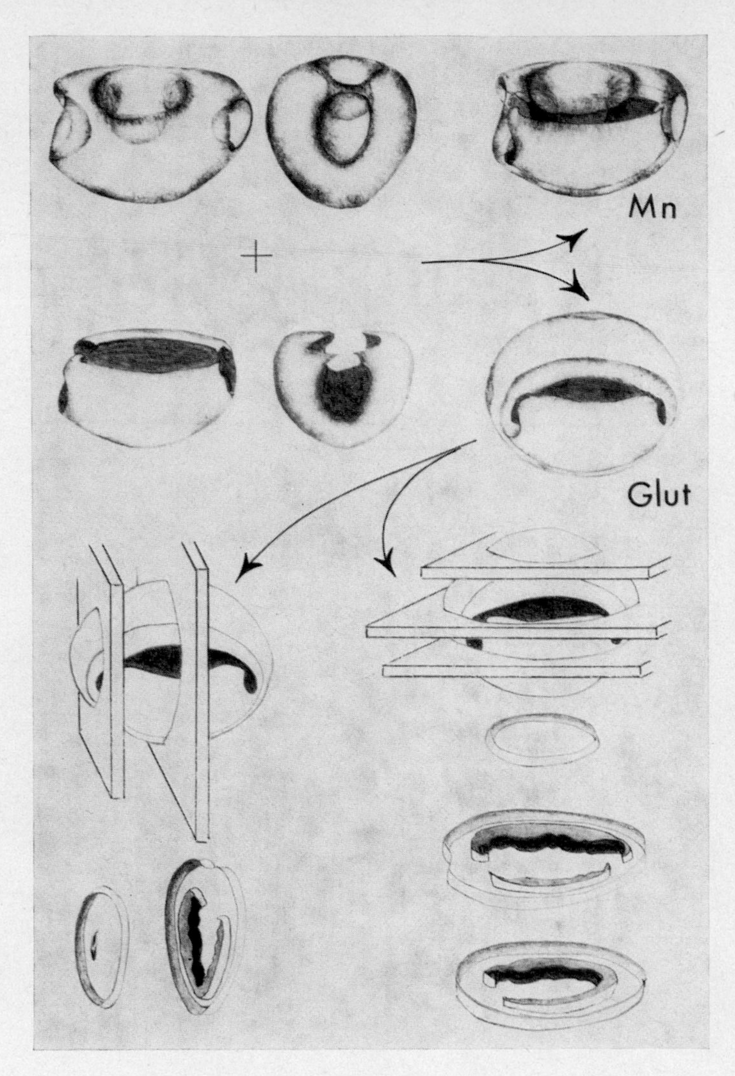

Fig. 10. The three-dimensional structure of a gamma particle (see text for explanation)

IV. Decay of Gamma Particles

Meanwhile, we had also turned our attention to the question: what happens to gamma particles when a zoospore encysts and then germinates? Electron-microscopic studies [TRUESDELL and CANTINO, 1970,1971] of zoospores undergoing synchronous encystment led to the conclusion that the first detectable morphological changes in a zoospore after it has been induced to encyst involve vesiculation by two key membrane systems: one of them is the *undifferentiated* portion of the backing membrane (fig. 11), the other is the gamma particle. A third change, not previously mentioned because our electron-microscopic evidence was still incomplete, but which [as suggested by HELD, 1972] deserves serious consideration, may be the disintegration of the microtubule triplets (fig. 7) that extend upward from the kinetosome along the outside of the nuclear apparatus. These triplets have never been seen in spores that have retracted their flagella [CANTINO and TRUESDELL, unpublished obs.].

Gamma particles decay in a very characteristic fashion during encystment. In brief (fig. 12, frames 1 and 2), after a zoospore has been triggered to encyst, the GS membrane starts to extend irregularly into the cytoplasm and begins to pinch off vesicles which then migrate to the plasma membrane. Simultaneously, the gamma matrix becomes highly idiomorphic as it produces a tier of vesicles 80 nm in diameter lined up in a regular fashion on its inner surface (fig. 13). These vesicles begin to fuse with the GS membrane (fig. 12, frame 3) in two ways. On the one hand, the GS membrane invaginates (upper gamma particle in frame 3) and seems to fuse with some of these vesicles. On the other hand, vesicles of 80 nm in diameter are simultaneously released from the gamma matrix, migrate to the GS membrane, and fuse with it. Since membrane material is apparently added to the GS membrane faster than it is lost, the GS membrane increases in size. As this process unfolds, the spore encysts (see fig. 14 for electron micrographs of a zoospore immediately after encystment) and the first cyst wall is laid down. Eventually (fig. 12, frame 4), these expanding gamma particles with decaying gamma matrices fuse with one another to yield large vesicles that contain two or more decaying

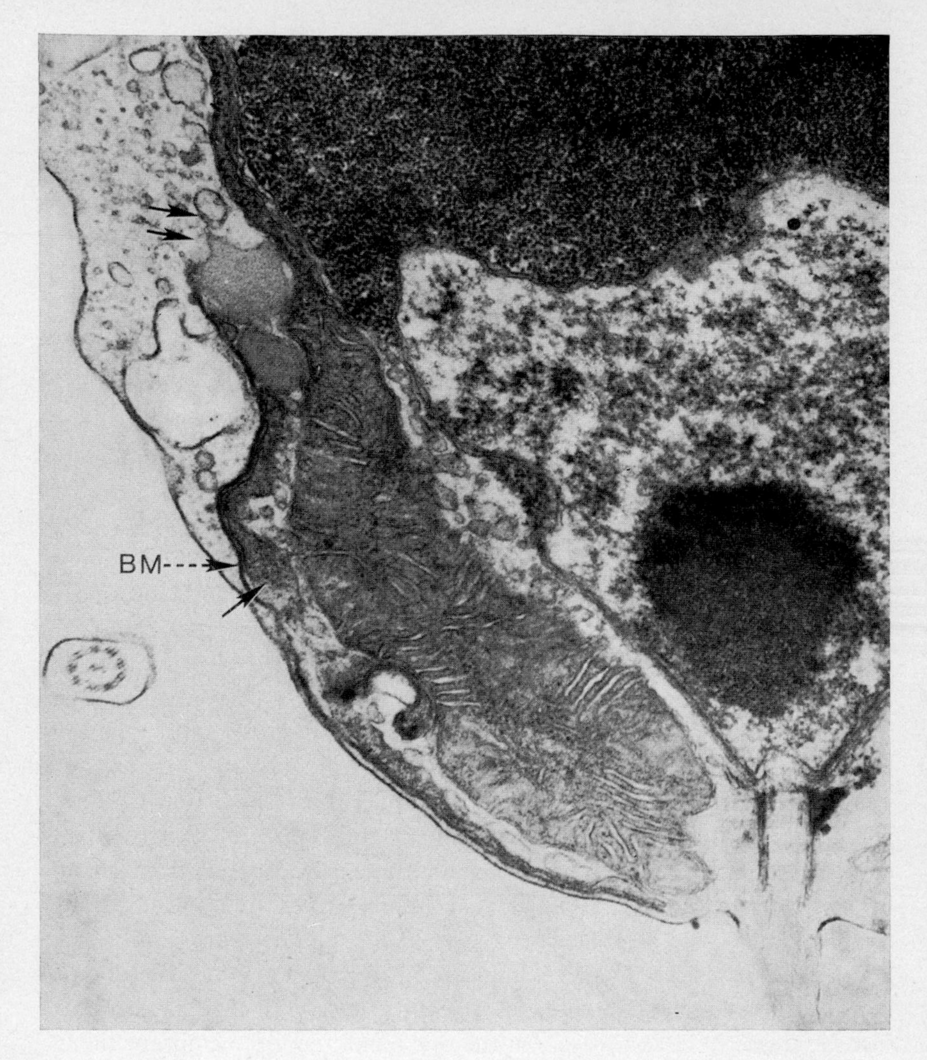

Fig. 11. One of the first detectable changes in a zoospore that has been triggered to encyst occurs in the backing membrane (BM). The latter consists of two portions: one of them (dotted arrow), which lies very close to the SB matrix (solid arrow), is 'differentiated' in that its two unit membranes are closer together than are the two members of other double membranes in the zoospore, and the region between them is partially occupied by an osmiophilic substance; the other part of the BM is undifferentiated, i.e., it looks like a typical double membrane, and it fragments rapidly (double arrow) just before encystment. Zoospore was prepared with fixation procedure 2 of TRUESDELL and CANTINO [1970].

matrices (fig. 15). All along, the GS membrane continues to release vesicles to the cell surface, the site of ongoing cell wall deposition. Finally, the decaying gamma particle matrices are entirely used up, leaving behind large multivesicular bodies which show no evidence that they once housed the cores of gamma particles.

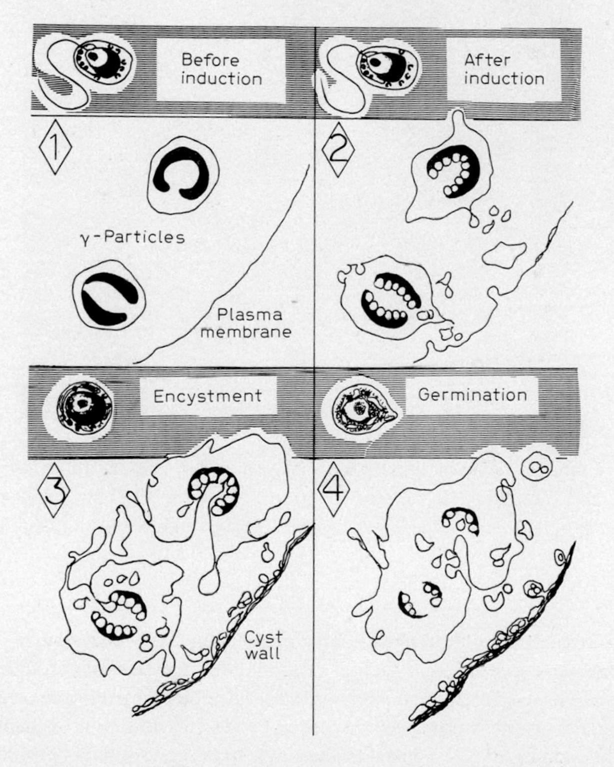

Fig. 12. The decay of gamma particles during zoospore encystment (see text for explanation).

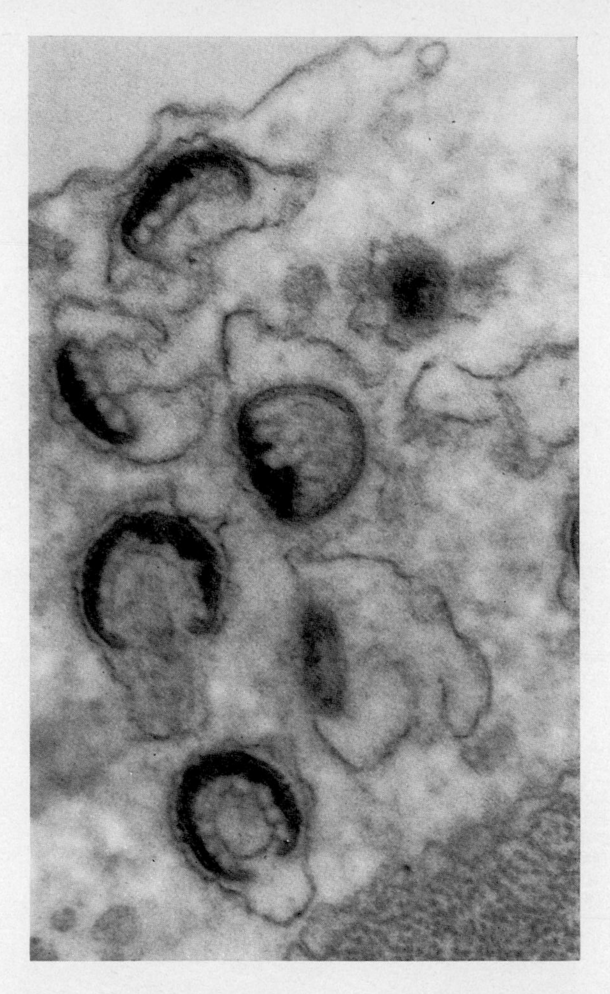

Fig. 13. Gamma particles fixed with OsO_4 and Ur acetate (fixation procedure 2, [TRUESDELL and CANTINO, 1970]) illustrating the presence of 80 nm vesicles on the inner surface of the gamma matrix. These vesicles are not seen in $KMnO_4$-fixed gamma particles.

Fig. 14. Cross sections through a spore immediately after encystment. The retracted flagellar axoneme is seen in cross section (X); longitudinal section through the retracted axoneme as seen in another spore is shown at (Y). The double membrane originally present around the nuclear cap has now been disrupted by vesiculation, thus exposing the mass of nuclear cap ribosomes to the cytoplasm. The single mitochondrion has partially subdivided itself to yield a many-armed yet single structure that typically spreads out over the ribosome mass at this stage in development [SOLL and SONNEBORN, 1971a; TRUESDELL and

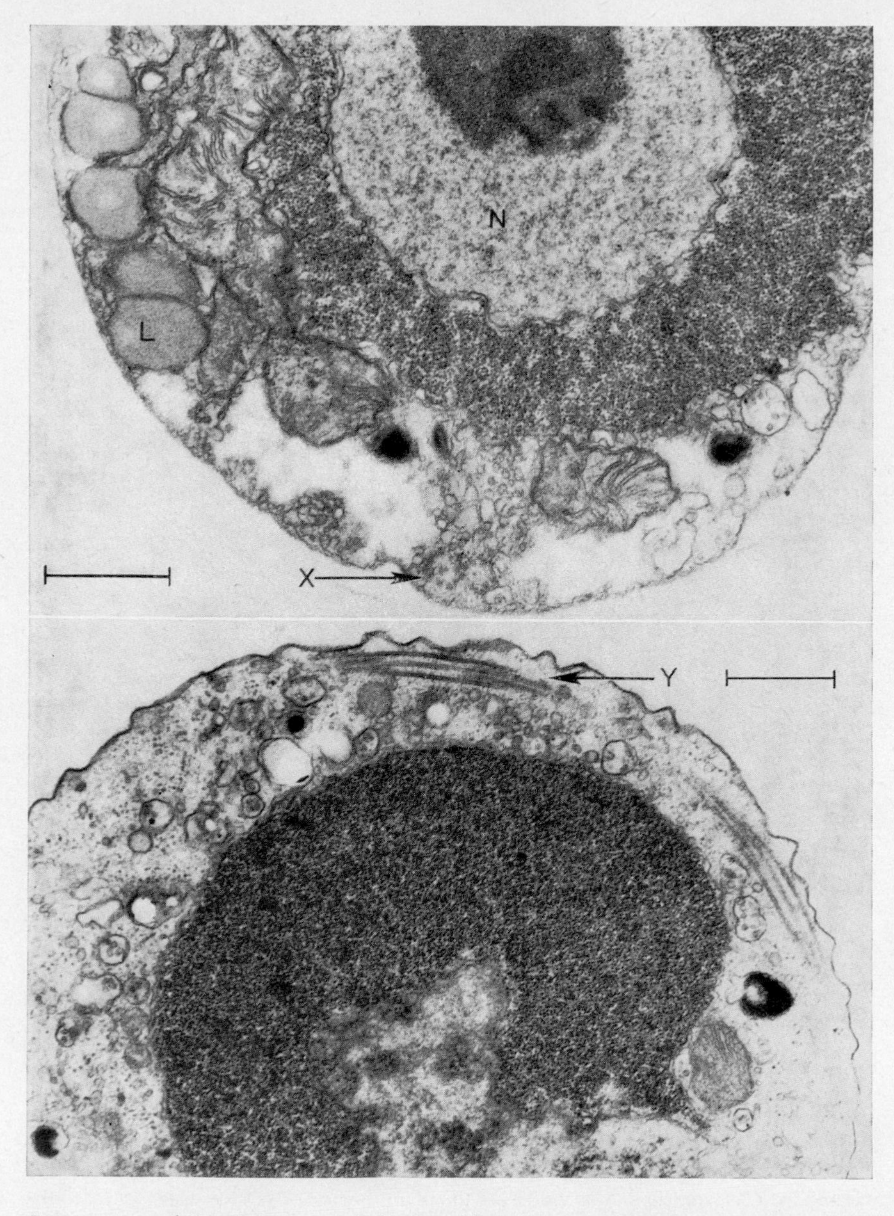

CANTINO, unpublished data]; about four profiles of it are visible here. The backing membrane originally around the lipid sac is disrupted by the time encystment is completed, but the lipid globules (L) still tend to remain clustered together at this stage. The nucleus (N) with its enclosed nucleolus remains membrane-bound. Prepared with fixation procedure 3 of TRUESDELL and CANTINO [1970]. Markers = 1 μm.

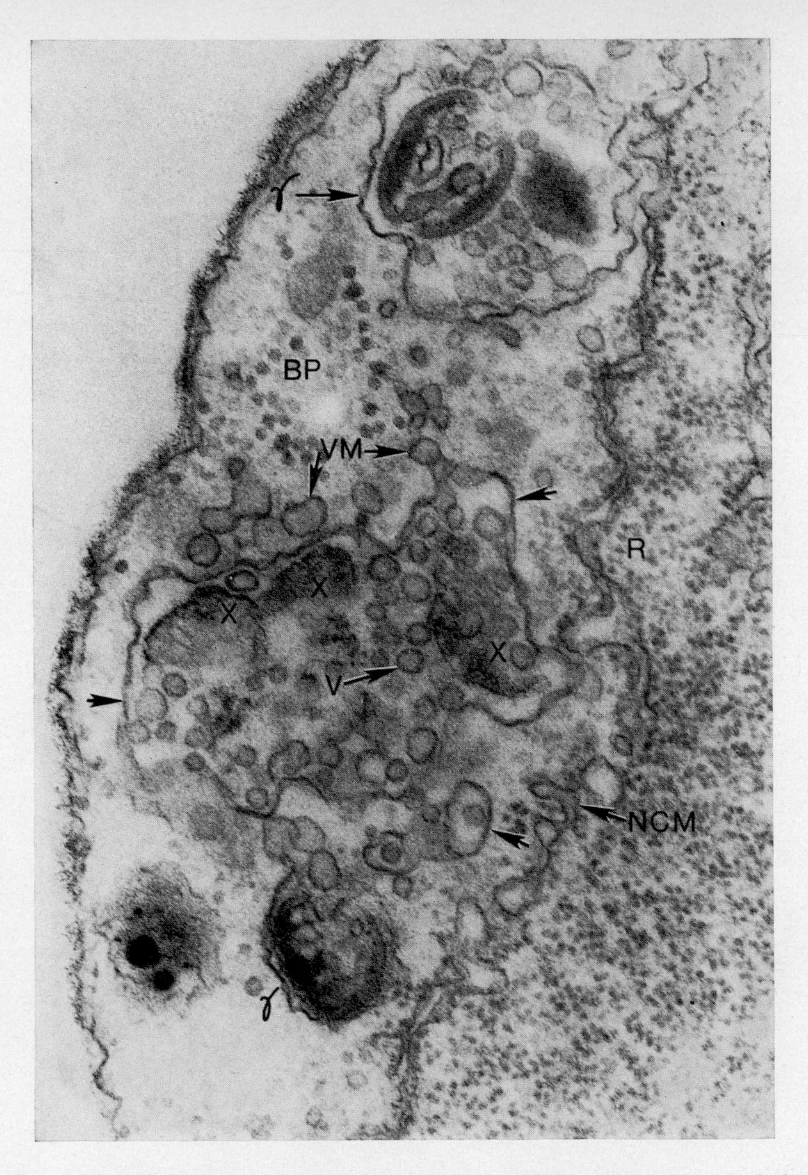

Fig. 15. Section through a large multivesicular body (MVB) resulting from the fusion of at least three decaying gamma particles. Arrowheads point to the continuous membrane that surrounds three decaying gamma matrices (X) and numerous 80 nm vesicles (V) released from the matrices. Also visible are the migrating vesicles (VM) which travel to the plasma membrane, where cyst wall material is deposited; two individual decaying gamma particles (γ), one of which may be in the process of fusing with the MVB; granules of *Blastocladiella* polysaccharide (BP); nuclear cap ribosomes (R); and vesiculation of the nuclear cap double membrane (NCM). Zoospore was fixed with OsO_4 and Ur acetate (fixation procedure 2 of TRUESDELL and CANTINO [1970].

V. Genesis of Gamma Particles

The sequence of morphological transformations leading to the formation of gamma particles during sporogenesis, first uncovered by LESSIE and LOVETT [1968], has now been re-examined in greater detail[3]. The salient features and interpretations of this process, as it occurs during the final 2 h of sporangium and zoospore differentiation, are summarized below.

The earliest identifiable progenitors of gamma particles are electron-dense granules (fig. 16a) which appear during papilla formation within swollen and irregularly flattened cisternae continuous with rough endoplasmic reticulum; these granular 'sub units' are 40 nm in diameter and have electron-transparent centers. They then increase in number and fuse together to yield doublet an triplet configurations (fig. 16b). At about this time, ribosomes are aggregating; eventually, the nuclear cap is formed in this way. Further coalescence among the granules of 40 nm in diameter, as well as the doublets and triplets, results in the formation of aggregates of 100 nm in diameter[4], which nearly fill the now fairly spherical smooth-surfaced cisternae (fig. 17a). These aggregates continue to fuse with one another to form larger – and hence, fewer – bodies ellipsoidal in shape and about 300 nm in diameter (fig. 17b); by rearrangement and coalescence, the components eventually take on the characteristic appearance of fully differentiated gamma matrices (fig. 17c). It appears as if the smooth cisternae, in which the 40 nm granules first appear, may be transformed into the final GS membranes. It is noteworthy

[3] We are exceedingly grateful to Dr. W. E. BARSTOW, Purdue University, for providing us with the unpublished electron micrographs shown here in figures 16 and 17, and an excerpt of his findings from which we have distilled this brief digest of the latest knowledge about the genesis of the gamma particle.

[4] These aggregates of 100 nm in diameter correspond to the dense 'granules' which were postulated by LESSIE and LOVETT [1968] to condense into the matrices of their 'vesicle-enclosed cup-shaped granules', and which TRUESDELL and CANTINO [1970] had estimated (from the pictures published by LESSIE and LOVETT) to be about the same dimensions as the 80 nm vesicular sub units released by the matrices of gamma particles (chapter IV) as they decay during zoospore encystment.

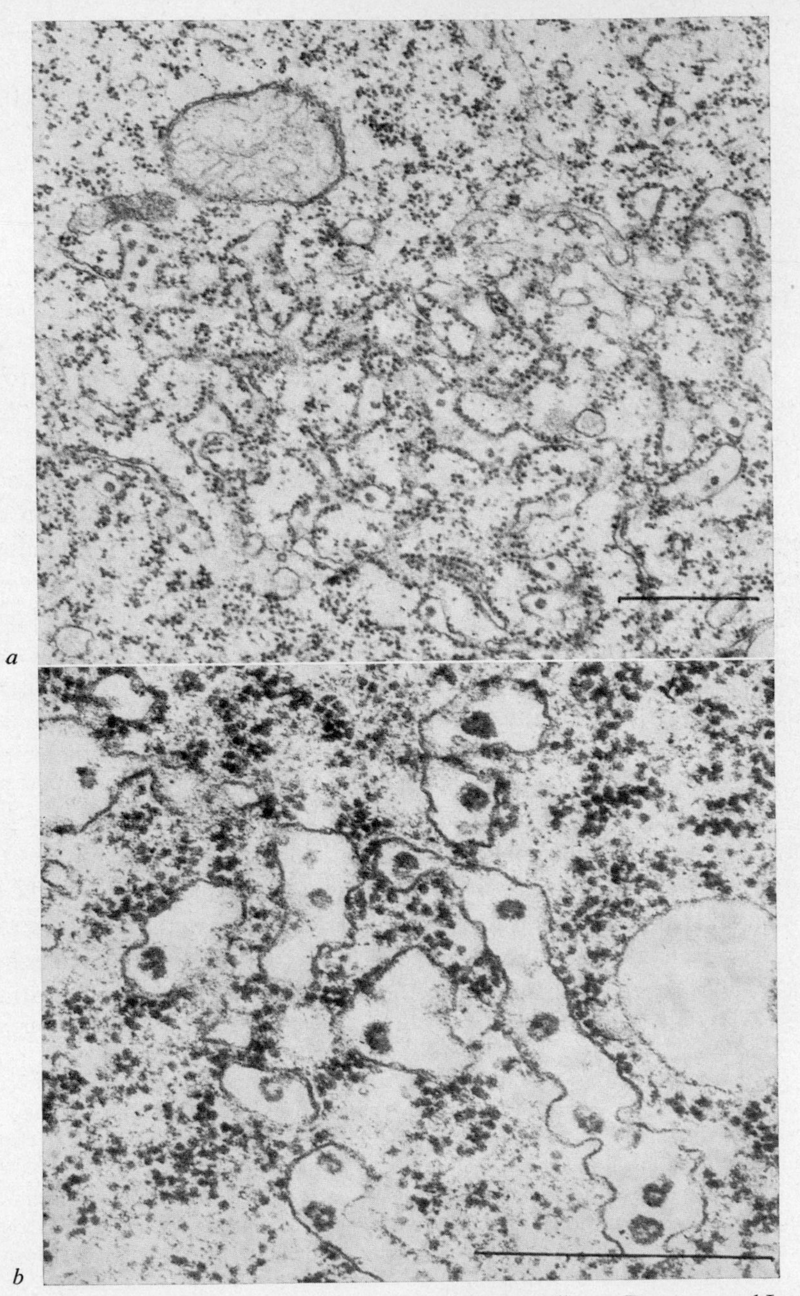

Fig. 16. Aspects of the genesis of gamma particles according to BARSTOW and LOVETT (personal communication) (see text for explanation). Markers = 0.5 μm.

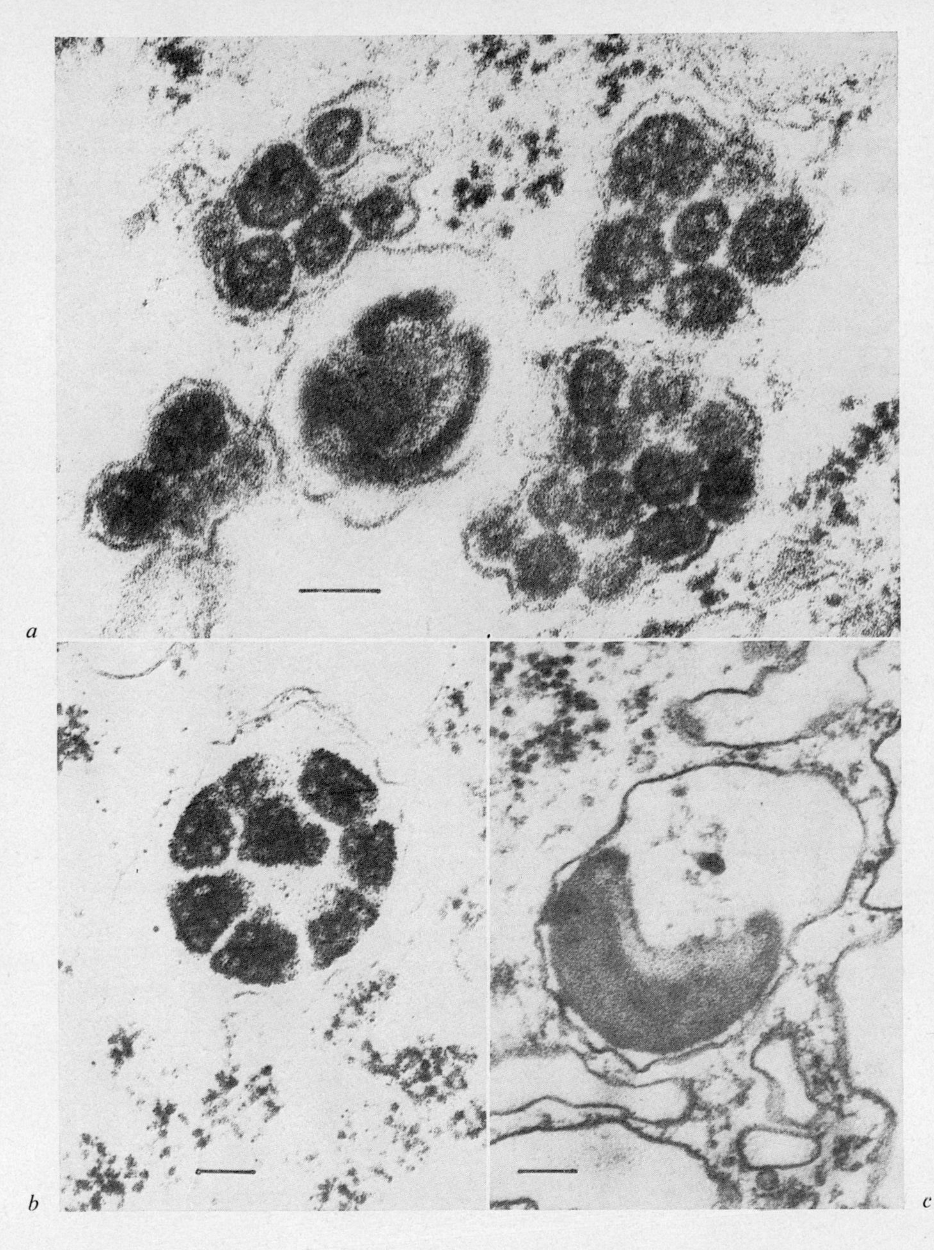

Fig. 17. Aspects of the genesis of gamma particles according to BARSTOW and LOVETT (personal communication) (see text for explanation). Markers = 0.1 μm.

that the stages described above for the genesis of a gamma particle in an OC plant also seem to occur during zoosporogenesis in a microcycle plant (chapter I; fig. 1). The fact that gamma particles apparently form only in cisternae of rough endoplasmic reticulum could mean that the genesis of one or more of the precursors of a gamma matrix requires a *de novo* synthesis of protein.

VI. Staining Properties of the Gamma Particle

To characterize gamma particles adequately, cell-free preparations were required. The only obvious feature that might be useful in attempts to isolate them was their osmiophylic nature and fairly uniform size. Additional work was done, therefore, to distinguish gamma particles cytochemically, and hopefully, thereby establish criteria with which they might be tracked during attempts to recover them from spore homogenates.

A. Culture Methods and Preparation of Zoospore Suspensions

New day-to-day working cultures of OC plants were started every month or two using zoospores derived from RS clones of wild type *B. emersonii*. Zoospores from OC plants were innoculated onto Difco Cantino PYG agar in Petri plates and incubated at 22 °C in the dark. Zoospores suspensions were then derived from the first-generation OC plants by flooding them with H_2O; for details and background on methodology, see CANTINO and LOVETT [1964], LOVETT [1967], and CANTINO *et al.* [1968].

B. Staining Methods

1. Lipid Stains

a) Orange G. Suspensions of zoospores were concentrated by centrifugation at 1,000*g* for 3 min at 22 °C, fixed in 4% formaldehyde containing 1% $CaCl_2$ at pH 7 for 24 h, applied to microscope slides, and dried. These cells were then exposed to a 1% solution of orange G in 0.1% acetic acid for 10–15 min at 22 °C, washed with citrate buffer at pH 3, dried, and mounted in lactophenol.

b) Nile blue. Zoospores were filtered, concentrated, fixed, and applied to microscope slides (all as for orange G) and then placed in a 0.02% solution of Nile blue for periods up to 30 sec at 22 °C, washed in 1% acetic acid, dried, and mounted in glycerol.

c) Sudan IV. Zoospores were fixed either with fumes of OsO_4 for 5 min or with 4% formaldehyde containing 1% $CaCl_2$ for 24 h, applied to microscope slides, and air dried. These cells were then covered with a saturated solution of Sudan IV in 70% ethanol for 10 min at 22 °C, washed in 50% ethanol, dried, and mounted in glycerol.

2. Vital Stains

a) Neutral red. Zoospores were placed on microscope slides and mixed with a 0.1% solution of neutral red at 22 °C.

b) Janus green B, brilliant cresyl blue, benzopurpurin 4B, trypan blue, methylene blue, and brilliant vital red. The first three dyes were prepared as 0.1% solutions and the last three as 1.0% solutions; they were mixed with zoospores as for neutral red.

3. Deoxyribonucleic Acid (DNA) Stains

a) May-Grunwald-Giemsa. All operations were done at 22 °C. Zoospores were filtered and centrifuged (as for orange G), fixed in 2% glutaraldehyde containing 0.1 M Na-cacodylate-HCl buffer at pH 7.2 for 2 min, concentrated by centrifugation at 1,000*g* for 3 min, suspended in 2% buffered glutaraldehyde again, transferred to microscope slides, and dried. The cells were then immersed in 0.25% eosin and 0.25% methylene blue in methanol for 10 min, transferred to a $^1/_{15}$ strength Giemsa stain for 20 min, dehydrated twice in acetone, about 3 min each, once in acetone-xylol (2:1) and once in acetone-xylol (1:2), cleared in xylol for 10 min, dried, and mounted in glycerol.

b) Feulgen. Zoospores were filtered and concentrated (as for orange G), suspended in 0.5 ml H_2O, fixed in 0.5 ml 2% formalin containing 33 mM NaH_2PO_4 + 46 mM Na_2HPO_4, pH 7, for 50 min at 22 °C, transferred to microscope slides and dried. They were rinsed in H_2O and then digested with either: (a) Sigma pancreatic ribonuclease (RNAse; 1 mg/ml in 20 mM tris(hydroxy methyl)aminomethane (Tris), 45 mM $MgCl_2$, and 5 mM $CaCl_2$, pH 7.3) for 1–4 h at 40 °C; (b) Sigma deoxyribonuclease (DNAse; 0.2 mg/ml of the above RNAse buffer solution) for 1–4 h at 22 °C; (c) a combination of (a) and (b) for 2.5 h at 40 °C; or (d) 1 N HCl for 10–20 min at 22–55 °C. The cells were rinsed in H_2O and then exposed to Schiff's fuchsin-sulfurous acid reagent, adjusted to pH 3.0 with NaOH, according to DUTT [1963] for 45–75 min at 22 °C, rinsed thrice for about 3 min each in sulfurous acid, washed in H_2O, dehydrated in 95 and 100% ethanol, about 10 min each, cleared in xylol, and mounted in glycerol.

c) Azure A. Zoospores were filtered, concentrated, suspended, fixed, applied to microscope slides, and hydrolyzed, identically to Feulgen stains. After being rinsed in H_2O, cells were immersed in 0.5% azure A in 50 mM HCl containing 50 mM $NaHSO_3$ (adjusted to pH 3.0 as for Feulgen) for 1.25 h at 22 °C, rinsed in H_2O, dehydrated in 95 and 100% ethanol for about 10 min each, cleared with xylol, and mounted in glycerol.

4. Other Stains

a) Acid fuchsin. Zoospores were filtered and concentrated like orange G, fixed in 4% formaldehyde containing 1% $CaCl_2$ at pH 7 for 30 min, applied to microscope slides and dried. The cells were then exposed to 0.05% acid fuchsin, washed in H_2O, dried, and mounted in glycerol.

b) Carmine. Zoospores were prepared like acid fuchsin, placed in 0.025% carmine, washed in H_2O, dried and mounted in glycerol.

c) Eosin Y. Zoospores were prepared for staining like the May-Grunewald-Giemsa stain, exposed to 0.25% eosin Y in methanol for 30 sec at 22 °C, dehydrated (as for May-Grunewald-Giemsa), cleared in xylol, dried, and mounted in glycerol.

d) Alcoholic methylene blue. Zoospores, prepared for staining like the May-Grunewald-Giemsa stain, were exposed to 0.25% methylene blue in methanol for 30 sec at 22 °C, rinsed in H_2O, dried, and mounted in glycerol.

Table I. The reactions of zoospore components to various stains

Reagent	Color					
	gamma particle	nuclear cap	nucleus	lipid sac	mito-chon-drion	cyto-plasm
Lipid stains						
Sudan IV	–	–	–	red-brown	–	–
Orange G	yellow	–	–	–	–	–
Nile blue	dark blue	dark blue	dark blue	–	–	–
Nucleic acid stains						
Eosin Y	red	–	–	–	–	pink
Brilliant cresyl blue	blue	dark blue	dark blue (nucleolus)	–	–	–
Methylene blue (alcohol)	dark blue	dark blue	blue (incl. nucleolus)	–	–	–
May-Grunwald-Giemsa	pink	dark blue	pink (nucleolus, blue)	–	–	–
Feulgen	pink	–	pink	–	–	–
Azure A	blue	–	dark blue	–	–	–
Miscellaneous dyes						
Janus green B	–	–	–	–	blue	–
Neutral red	red	pink	–	–	–	–
Benzopurpurin 4B	–	–	red	–	–	–
Trypan blue	–	–	–	–	–	–
Methylene blue	–	blue	blue	–	–	–
Brilliant vital red	–	–	–	–	–	–
Acid fuchsin	–	–	–	–	–	red
Carmine	–	–	–	–	–	red

C. Results

1. Probable Association of Lipids with the Gamma Particle

Osmium tetroxide, which reacts [JENSEN, 1962] with unsaturated lipids forming black deposits, turned gamma particles very dark. Their possibly lipoidal nature was therefore examined further using lipid stains with different specificities (table I). Sudan IV, a neutral lipid indicator, did not stain gamma particles. The only structures consistently stained by it were components of the SB-lipid complex; they became red-brown. Gamma particles were colored blue by nile blue, a response typical of free fatty acids and phospholipids [JENSEN, 1962]. Nile blue acts as a basic dye; presumably the nucleus and RNA-rich nuclear cap were also stained blue for this reason. Orange G, a dye reportedly [JENSEN, 1962] specific for phospholipids, turned gamma particles yellow. The foregoing results suggest that lipid – and in particular phospholipid – is an integral component of the gamma particle.

2. Suggestive Evidence that the Gamma Particle Contains Nucleic Acid

Gamma particles were tested for their reaction to several nucleic acid stains (table I). General nucleic acid dyes consistently gave positive reactions. Eosin Y stained the zoospore cytoplasm light red and gamma particles a more intense red. Gamma particles reacted positively to both brilliant cresyl blue and methylene blue, as did the nuclear cap and nucleolus. The May-Grunwald-Giemsa stain reacted differentially with the 'nuclear apparatus' (i.e. the nucleus-nuclear cap combination) and gamma particles; the RNA-rich nuclear cap and nucleolus were colored blue, while gamma particles and the nucleus were colored pink. These results suggested that gamma particles might carry DNA; the Feulgen and azure A tests were therefore applied.

3. Selective Hydrolysis of Nucleic Acid and the Staining Characteristics of the Gamma Particle

The zoospore of *B. emersonii* has some convenient built-in controls for doing selective nucleic acid staining. Its massive nuclear cap contains ribosomes, and therefore, RNA; hence it can serve as an indicatoı for judging the effectiveness of any RNA hydrolysis. Similarly, its conspicuous nucleus can serve as an internal standard for estimating both the specificity of DNA staining techniques and the selectivity of any DNA hydrolysis. In other organisms, the time of hydrolysis required for maximal color intensity can differ among various tissues [SWIFT, 1955]. The timing of acid hydrolysis of zoospores also turned out to be important for maximum staining of gamma particles with both Feulgen and azure A tests (table II). Hydrolysis in 1 N HCl at 55°C for less than 11 min was not sufficient to yield visible color in gamma particles treated with either the azure A or the Feulgen technique. The nuclei did stain, but less so than under optimal conditions. Gamma particles became visibly colored after 11 min of hydrolysis, but to maximum intensity only after 18–20 min and 15–18 min of hydrolysis for the azure A and Feulgen tests, respectively. These time intervals were also the ones that yielded maximum nuclear staining. Complete hydrolysis of RNA and elimination of residual staining of the nuclear cap by azure A required 11 min.

Table II. Effect of duration of HCl hydrolysis on the response of gamma particles, nuclei, and nuclear caps to the azure A and Feulgen staining procedures

Duration of hydrolysis[1], min	Staining intensity[2]		
	gamma particles	nucleus	nuclear cap
Azure A			
5	–	++	+
7	–	++	+
9	–	++	+
11	+	+++	–
15	++	++++	–
18	+++	++++	–
20	+++	++++	–
Feulgen			
5	–	+	–
7	–	++	–
9	–	++	–
11	+	+++	–
15	++	+++	–
18	++	+++	–

[1] Hydrolysis in 1 N HCl at 55 °C. Hydrolysis in 5 N HCl at 23 °C was not effective in eliminating RNA.

[2] No detectable staining(–); most intense staining (++++).

Selective enzymatic digestion of RNA and DNA provided additional insight into the nucleic acid content of gamma particles (table III). Following hydrolysis of zoospores with RNAse for 1–4 h, gamma particles appeared to be more intensely stained than they were in untreated control spores. A 2–3 h treatment with DNAse eliminated the stainability of both gamma particles and nuclei with azure A and Feulgen reagents. These results suggested that gamma particles contained DNA; the fact that they appeared to stain more intensely for DNA following removal of RNA suggested that RNA may also have been present.

Under optimal conditions for staining, not all the spores contained distinctly colored gamma particles; furthermore, not all of the gamma particles in any one spore were always stained. The best results were obtained with azure A; this may have been because DNA is three to four times more sensitive to azure A than to Schiff's reagent [Swift, 1962].

Table III. Effect of selective enzymatic hydrolysis of nucleic acids on the staining characteristics of gamma particles, nuclei, and nuclear caps

Enzyme	Duration of treatment, h	Intensity[1] gamma particles	nucleus	nuclear cap
Feulgen[2]				
RNAse	1	++	+++	–
	2	++	+++	–
	3	++	+++	–
	4	++	+++	–
	control	+	+++	+++
DNAse	1	+	+	+++
	2	–	+	+++
	3	–	–	+++
	4	–	–	+++
	control	+	+++	+++
Azure A[3]				
RNAse	2	+++	++++	–
	2.5	+++	++++	–
	control	++	++++	+++
DNAse	2	–	–	+++
	2.5	–	–	+++
	control	++	++++	+++

[1] No detectable staining (–); most intense staining (++++).
[2] Adjusted to pH 2.8 for Feulgen staining.
[3] Adjusted to pH 3 for azure A staining.

4. Other Staining Reactions in B. emersonii *Zoospores*

In living spores, Janus green B stained mitochondria blue; trypan blue and brilliant vital red apparently did not react with any intracellular component; and methylene blue colored nuclear caps and nuclei blue (table I). In fixed spores, acid fuchsin and carmine stained the cytoplasm red; the same color was obtained when nuclei were stained with benzopurpurin 4B. The only other compound that consistently stained gamma particles was neutral red (table I); it turned the gamma particles dark red and nuclear caps pink.

D. Discussion

1. Association of Lipids with the Gamma Particle

Osmium tetroxide forms a black deposit in the presence of unsaturated fats, as well as proteins. The gamma matrix, which seems to contain, at least during its decay, a very compact system of membranous elements [TRUESDELL and CANTINO, 1970], turned black in its presence. A variety of techniques involving OsO_4 [JENSEN, 1962] have been used for the cytochemical demonstration of complex lipids, including phospholipids in myelin sheaths, Golgi elements, red-cell envelopes, etc. Gamma particles reacted positively to two phospholipid dyes, but they had no apparent affinity for those used to stain neutral lipids. Therefore, it seems likely[5] that lipid is associated with the gamma particle, especially with its differentiated core, and that it is probably complex and contains phospholipid.

2. Use of Neutral Red in Attempts to Isolate Gamma Particles

The selective staining of gamma particles with neutral red was probably the most practical result of the foregoing work. The small size of the gamma particle – it being barely resolvable with the light microscope – virtually ensured that other characteristics would be needed in any initial attempts to track these organelles during fractionation of spore homogenates. When used as a vital dye, neutral red was quickly absorbed by gamma particles, imparting to them a bright red hue which was retained even after the zoospores had lysed. This meant that gamma particles could be followed microscopically during attempts to isolate them.

[5] Some experiments have now been done to determine the lipid composition of gamma particles. The results indicate that it is composed of 28.3% neutral lipids, 37.7% phospholipids, and 33.9% glycolipids. Major components of the neutral lipids are sterols and sterol esters with some mono-, di-, and triglycerides present. The phospholipid fraction consists of 4.9% phophatidic acid, 17.7% phosphatidyl ethanolamine, 10.4% lysophosphatidyl ethanolamine, 52.3% phosphatidyl choline and lysophosphatidyl choline, 9.8% phosphatidyl inositol, and 4.9% phosphatidyl serine. Gamma particle glycolipids contain one major component and at least three minor components. The protein: lipid ratio in gamma particles is about 1:2.

VII. Isolation of Gamma Particles

Several characteristics of gamma particles were useful in designing the methods to be employed for their isolation. First, they lie free in the cytoplasm and are not confined by the backing membrane or connected to the flagellar apparatus, in contrast to all other well-defined organelles (fig. 6) in the spore. Second, gamma particles are remarkably uniform in size, in the mean number per spore (provided that lighting conditions are controlled and that spores are not allowed to swim for extended periods of time; see chapter XI), and in their frequency distribution patterns (fig. 4) among the motile cells produced by the various phenotypes of *B. emersonii*. Third, gamma particles have a strong affinity for neutral red when it is used as a vital dye (chapter VI, section C, 4). And last, gamma particles have an intensely electron-opaque matrix with a very characteristic morphology which permits their easy identification in electron micrographs.

The purpose of the work to be described below was: (1) to develop a medium that would maintain the integrity of gamma particles during isolation; (2) to perfect an efficient means of lysing spores which would bring about release of gamma particles while simultaneously minimizing fragmentation of other cell structures; (3) to devise a scheme for differential centrifugation of spore lysates which would yield a fraction greatly enriched in gamma particles, and (4) to establish procedures for the further purification of such enriched fractions via sucrose density gradient centrifugation.

Some of the observations described below led us to eliminate certain kinds of procedures from further consideration as we worked out the method eventually adopted to isolate gamma particles. These experiments and results are nonetheless provided here in some detail, because we think that they can be useful to those who may try to improve our methodology or attempt to isolate gamma particles under different environmetal conditions. However, we urge others to go directly to procedure E (this section, A, 5) in which the method we finally adopted for the isolation and purification of gamma particles is described.

A. Methods and Materials

Unless otherwise stated, OC plants were grown in Petri plates on PYG agar at 22 °C in the dark, all zoospore suspensions were filtered through either Whatman No. 1 or Sargent No. 500 paper to remove germlings, and all centrifugation operations were performed at room temperature, about 23 °C.

1. Procedure A: Tracking Gamma Particles with Neutral Red

a) Culture methods and preparation of zoospore suspensions. To prepare zoospore suspensions, one culture of a suitably long series of plates was flooded with 10 ml of one of the following: H_2O; 1 mM phosphate (Na) + 1 mM $CaCl_2$ at pH 6; 1 mM phosphate (Na) + 0.1 mM $CaCl_2$ at either pH 6 or 8.5. The plate was swirled for about 2 min and the resulting zoospore suspension was then transferred to a second culture which was manipulated as before. This process was repeated up to four times for every plate in the series, the final zoospore suspension from the last culture of each series being pooled with previous final suspensions. The pooled suspension containing about 2×10^8 spores was either chilled at 0–4 °C for 5–10 min or used without chilling.

b) Staining of zoospore suspensions with neutral red. Zoospore suspensions were concentrated by centrifugation for 3 min at 3,900 *g* and then either: (a) resuspended for 2–3 min in 10 ml of staining solution (1 µg neutral red/ml 1 mM Na_2HPO_4 + 1 mM $CaCl_2$) at 0–4 °C; (b) resuspended in 10 ml of 0.25 M sucrose, washed by centrifugation for 3 min at 3,900 *g*, and then suspended again in 10 ml of an alternate staining solution (5 µg neutral red/ml 0.25 M sucrose); (c) chilled and stained immediately in 50 µg neutral red/ml H_2O.

c) Production of zoospore Homogenates. Zoospores containing visibly stained gamma particles were washed by centrifugation for 3–5 min at 1,000–5,400 *g* and then resuspended in one of the following solutions at 0–4 °C: 10 ml of 0.25 M sucrose, pH 7.6; 1 ml of 1 mM phosphate (Na) + 1 mM $CaCl_2$, pH 7.6; 1 ml of 1 mM phosphate (Na) + 0.1 mM $CaCl_2$, pH 8.5. Spores suspended in the latter two solutions were then either disrupted directly with a glass homogenizer for 15 min at 0–4 °C, or first washed by centrifugation for 3–5 min at 2,000–5,400 *g*, resuspended in 1 ml of 1 mM Na_2HPO_4 + 1 mM $CaCl_2$, pH 8.5, and then disrupted either with a glass homogenizer or by lysis with 0.05% Na lauryl sulfate (SLS) at 0–4 °C. Spores suspended in 0.25 M sucrose were washed by centrifugation for 3–5 min at 2,000 *g* and then osmotically disrupted by resuspension in 10 ml of 1 mM phosphate (Na) at pH 6.

d) Differential centrifugation of zoospore homogenates. Zoospore homogenates were diluted 1:10 with either 0.25 M sucrose at pH 7.6, 1 mM phosphate (Na) + 1 mM $CaCl_2$ at pH 7.6, or 1 mM phosphate (Na) + 0.1 mM $CaCl_2$ at pH 6, and subjected to differential centrifugation: 3,900 and 5,400 *g* for 3 min each and 6,600 *g* for 15–20 min for the first two media; 1,200, 2,600 and 3,900 *g* for 3 min each, and then 5,400 and 8,600 *g* for 10 min each for the final suspending medium. Pellets and supernatants were examined by phase and light microscopy.

2. Procedure B: Establishing the Cause of Gamma Particle Aggregation

a) Preparation of zoospore suspensions. Zoospore suspensions were produced by successively flooding OC thalli as in procedure A (this chapter, section A, 1, a) using 10 ml of either 1 mM phosphate (Na) at pH 7.6, unbuffered 3 mM $MgCl_2$, or H_2O. The combined zoospore suspension resulting from four successive floodings, containing about 10^8 spores, was then washed by centrifugation at 1,000 *g* for 3 min, and resuspended in either 1 ml of H_2O, 1 ml of 3 mM $MgCl_2$, 10 ml of 0.05% SLS, or 5 ml of 5 mM ethylenediamine tetraacetic acid (EDTA) at pH 8.3; in some instances, the spores were chilled at 0–4 °C for 5 min before washing.

b) Production of zoospore homogenates. Zoospores suspended in SLS lysed on contact. Zoospores suspended in $MgCl_2$ or H_2O were disrupted in a glass homogenizer at 0–4 °C for 15 min. Zoospores suspended in EDTA were homogenized by repeated ejections through a 27 gauge needle according to LOVETT [1963].

c) Differential centrifugation of zoospore homogenates. The homogenates in $MgCl_2$ or H_2O were diluted 1:10 with either 3 mM $MgCl_2$ or H_2O, and subjected to differential centrifugation: 1,000, 1,600, 2,600 and 3,900 *g* for 5 min each. Spores lysed in SLS and spores homogenized in EDTA were centrifuged without further dilution: 1,600, 7,000, and 9,000 *g* for 5 min each (for SLS), and 1,000 *g* for 10 min, 7,000 *g* for 5 min, and 14,000 *g* for 20 min (for EDTA). Pellets and supernatents were examined by phase and light microscopy.

3. Procedure C: Isolation of Gamma Particles by Differential Centrifugation

a) Zoospore suspensions. Zoospore suspensions were produced by successively flooding OC thalli as in procedure A (this chapter, section A, 1, a) but using only H_2O for the purpose. The suspensions were chilled, stained with 0.005% neutral red for 5 min at 0–4 °C, concentrated by centrifugation for 3 min at 1,000 *g* and 0–4 °C, resuspended in 20 ml of 0.3 M sucrose containing 1 mM KCl, 10 mM Tris, and 0.2 mM EDTA at pH 7.8, washed by centrifugation for 3 min at 1,000 *g* and 0–4 °C and finally resuspended in 2 ml of the same solution.

b) Production of zoospore homogenates. The suspensions, which contained about 2×10^8–4×10^8 spores, were homogenized as in procedure B (this chapter, section A, 2, b).

c) Differential centrifugation of zoospore homogenates. The homogenates were diluted 1:4 with 0.3 M sucrose containing 10 mM KCl, 10 mM Tris, and 0.2 mM EDTA at pH 7.8 and subjected to differential centrifugation: 750 *g* for 10 min, 7,000 *g* for 5 min, 10,000 *g* for 20 min, and 54,000 *g* for 20 min, all at 0–5 °C. Fractions were examined by phase and light microscopy.

4. Procedure D: Determination of Optimal Conditions for Gamma Particle Isolation

a) Differential centrifugation of zoospore homogenates. Zoospore suspensions containing 2×10^8–9×10^8 spores were prepared and homogenized as in procedure C (this chapter, section A, 3, a, b) but without staining, in one of the following solutions: medium A (10 mM KCl, 10 mM Tris, and 0.2 mM EDTA in 0.3 M sucrose at pH 6.5); medium B (10 mM Tris and 0.2 mM EDTA in 0.3 M sucrose at pH 6.9); medium C (10 mM KCl, 10 mM Tris, and 0.2 mM EDTA in 0.4 M sucrose at pH 6.9); medium D (10 mM KCl, 10 mM Tris, and 0.2 mM EDTA in 0.3 M sucrose at pH 7.4); medium E (10 mM KCl, 10 mM Tris, and 0.2 mM EDTA in 0.3 M sucrose at pH 8.5); medium F (10 mM KCl, 10 mM phosphate [Na], and 2 mM EDTA in 0.4 M sucrose at pH 6.5). The zoospore homogenates were diluted 1:8

with medium A, B, C, D, E, or F, respectively, and subjected to differential centrifugation as illustrated in flow sheet 1.

5. Procedure E:
Final Procedure for Isolation and Purification of Gamma Particles

a) Preparation of zoospore suspensions. Zoospore suspensions were produced by flooding single generation cultures of OC plants with 5–6 ml H_2O. After intermittent swirling for 10–20 min, the spore suspensions from each plate were individually filtered, i.e. they were not used for series flooding as in procedure A (this chapter, section A, 1, a).

b) Production of zoospore lysates. The zoospore suspensions containing between 10^9 and 1.5×10^{10} spores were pelletd at 1,000 *g* and 0–4 °C for 3–5 min, resuspended in 8–12 ml of 1 M sucrose containing 25 mM phosphate (K), 1 mM KCl, and 0.5–1 mM ethyleneglycol-bis (β-aminoethyl ether)-N,N′-tetraacetic acid (EGTA) at pH 6.8, and held at 0–4 °C for 5–10 min. The zoospores were then osmotically lysed by replacing the spent 1 M sucrose solution (by centrifugation) with either 8 ml of 0.4 M sucrose containing 25 mM phosphate (K), 1 mM KCl, and 0.5–1.0 mM EGTA at pH 6.8 or 4 ml of 25 mM phosphate (K) containing 1 mM KCl and 0.5 mM EGTA. Lysis was complete within 5 min at 0–4 °C, judging by the loss of cell refractility as seen with the phase microscope.

c) Differential centrifugation of zoospore lysates. The lysates were centrifuged for 5 min at 1,000 *g* and 0–4 °C, and the supernatants were combined with the spent 1 M supernatants from the zoospores (this chapter, section A, 5, b). These were centrifuged twice, first at 5,000 *g* and 0–4 °C for 5 min, and then at 20,000 *g* and 0–4 °C for 15–20 min. The 20,000 *g* pellets were enriched for gamma particles.

d) Purification of Suspensions of Gamma Particles. The foregoing gamma particle enriched pellets were further purified by resuspension in 1 M sucrose containing 25 mM phosphate (K), 1 mM KCl and 0.5 mM EGTA at pH 6.8, and then subjecting them to either discontinuous (2.5 M sucrose, 1 ml; 2.25 M, 2.02 M, 1.8 M and 1.58 M sucrose, 1.5 ml each; 1.14 M sucrose, 1 ml; prepared in 25 mM phosphate containing 10 mM KCl and 0.5–1 mM EGTA, pH 6.8) or isopycnic (2.5–1.3 M sucrose prepared in the same phosphate buffer) sucrose density gradient centrifugation at 0–4 °C. In this and previous sections, centrifugations requiring forces under 5,000 *g* were made with an International model HR-I refrigerated centrifuge; otherwise, an International model BD-2 refrigerated vacuum centrifuge was employed using the six-place No. 969 swinging-bucket head.

e) Electron microscopy of gamma particles. The gamma particle preparations were fixed in a cacodylate buffered solution of OsO_4 and glutaraldehyde [TRUESDELL and CANTINO, 1970] and processed for electron microscopy according to CANTINO and MACK [1969].

B. Results

1. Production of Zoospore Lysates

The zoospore's plasma membrane can be ruptured very easily by any one of several methods. The technique employed depends, in part at least, upon the medium in which the cells have been suspended.

Zoospores suspended in hypotonic or nearly isotonic solutions can be disrupted with a tight fitting glass homogenizer. The spore lysates thus produced contained mainly undefinable 'vesicle-like' bodies; most of the easily recognizable organelles in the spore (fig. 6) had been either destroyed or so drastically altered in morphology that they were no longer distinguishable. Furthermore, this method proved to be impractical with the more viscous hypertonic sucrose solutions. For these reasons, alternative means for producing spore lysates were examined.

LOVETT's [1963] technique for lysing zoospores of *B. emersonii* enabled him to isolate nuclear caps free of nuclei and mitochondria; spores suspended in a hypotonic solution were ruptured by repeated extrusion through a hypodermic needle. When we applied this technique (this chapter, section A, 2, b), the resulting cell lysates looked much like those obtained using glass homogenizers. Increasing the tonicity of homogenizing media with sucrose lessened the degree of cell fragmentation, but the technique now required a prolonged, 30 min treatment; to obtain 90% lysis, 25–30 extrusions through the needle were needed. The presence of intact mitochondria, nuclei, and nuclear caps in the lysates was evidence that this approach was less severe than that utilizing glass homogenizers. This technique was satisfactory when relatively low numbers of spores were used, but it proved to be impractical when higher population densities (more than approximately 10^9 spores/ml) were required. Very dense suspensions were extruded through the needle only with great difficulty, especially at low temperatures such as 0–4 °C. As a result, considerable time was required to produce a suitable amount of spore disruption.

As mentioned earlier (fig. 6, legend), the major organelles in the zoospore – the nuclear apparatus, mitochondrion, SB matrix, and associated lipid globules –, seem to be held in position, at least in part, by the backing membrane (BM) which is composed of two parallel unit membranes. Gamma particles are the only prominent organelles not confined by the BM. It seemed as if it should be possible to osmotically rupture the plasma membrane while leaving the backing membrane intact. If achieved, this technique would have permitted ready separation of gamma particles from the other 'bound' organelles and thereby facilitated the isolation of gamma particles by lessening the amount of debris customarily produced with the previous homogenization techniques.

Differential rupturing (this chapter, section A, 5, c) of the plasma membrane while leaving the BM intact was accomplished by suspending zoospores in 1 M buffered sucrose at 0–4 °C for 5–10 min, pelleting them, and resuspending the pellet in 0.4 M buffered sucrose ad 0–4 °C. Spores lysed in this manner resembled living spores morphologically but they no longer possessed an intact plasma membrane judging by their loss of refractivity in the phase microscope. On the other hand, when osmotically shrunken spores were resuspended in phosphate, KCl, and EGTA without sucrose, *both* the plasma membrane and the BM were ruptured.

2. Tracking Gamma Particles with Neutral Red by Procedure A

Gamma particles in living zoospores turned red in about 2–3 min at 0–4°C in the presence of 1–50 μg/ml neutral red. They retained their red color even after the zoospores had lysed. The only other organelle stained under these conditions was the nuclear cap which, because it was much larger and less intensely stained was easily distinguished from gamma particles. Hence it seemed as if neutral red might serve as a useful marker dye during the initial attempts to isolate gamma particles from spore lysates.

Suspensions of spores which had been stained with neutral red and then homogenized and resuspended in a unbuffered 0.25 M sucrose solution were separated by differential centrifugation into three fractions. Centrifugation for 3 min at 3,900*g* produced large white pellets with small purple centers; they consisted mainly of unidentifiable large debris and unbroken spores. Centrifugation of the 3,900 *g* supernatants for 3 min at 5,400 *g* produced similar but smaller pellets with purple centers; these pellets contained mitochondria and smaller debris, but apparently no organelles of gamma particle size. Centrifugation of the 5,400 *g* supernatants for 15 min at 6,600 *g* produced very small purple pellets with a trace of overlying uncolored material. Many gamma particles were visible when this pellet was resuspended in the unbuffered sucrose solution; whenever these particles clung together in large aggregates, their red color was clearly evident. This fraction also contained lipid globules.

Zoospores that had been washed, stained with neutral red, and then homogenized in a Na phosphate buffer containing 1 mM $CaCl_2$ were fractionated by differential centrifugation. The only colored pellet containing gamma particles was the one resulting from centrifugation at 3,900 *g*; within it were aggregates of small red particles apparently adhering to cellular debris. Intermediate level centrifugations of the low-speed supernatants produced nearly colorless pellets consisting largely of nuclear caps, nuclei, and other large, undefinable debris. Very small amounts of material were pelleted at 6,600 *g*; phase-microscopic observations suggested that it consisted of vesicle-like bodies apparently smaller than gamma particles. These results implied that Ca may have caused gamma particles to be absorbed onto cellular constituents, thus forming large clusters by aggregation, and that these aggregates would sediment at lower centrifugal forces than the non-aggregated gamma particles in Ca-free media.

Lowering the $CaCl_2$ concentration to 0.1 mM and increasing the number of fractionation steps to five further emphasized the probable effect of Ca on gamma particles. Nuclear caps, nuclei, unbroken cells, and large debris were pelleted by centrifugation at 1,200 *g* for 3 min, thus producing a large white pellet with a red center. Centrifugation of the 1,200 *g* supernatant for 3 min at 2,600 *g* yielded a large red pellet containing lipid-SB matrix complexes (fig. 6) as well as cellular debris to which small, red-colored gamma particles were apparently adhering. In addition, red-colored aggregates of small vesicle-like bodies were evident. Most of the aggregated gamma particles were pelleted from the 2,600 *g* supernatant by centrifugation for 3 min at 3,900 *g*. This pellet was small and red. In contrast to the behavior of gamma particles found in homogenates produced in 1 mM $CaCl_2$, small red pellets were obtained following a 10 min, 5,400 *g* centrifugation of the 3,900 *g* supernatants and a 10 min, 8,600 *g* centrifugation of the 5,400 *g* supernatants. These pellets contained mainly gamma particles, singly and in small clusters. Therefore, lowering the $CaCl_2$ concentration ten-fold apparently decreased the tendency of gamma particles to adhere to cellular debris and to one another.

Concentrated zoospore suspensions in unbuffered 0.25 M sucrose solutions were osmotically lysed by resuspension in 1 mM Na phosphate. It was hoped that by rupturing the plasma membrane in this way, gamma particles would be released with minimal simultaneous cellular fragmentation. However, they did adhere to debris in the lysates.

The gamma particles which appeared to adhere to cellular debris in the 3,900 *g* pellets were released by resuspending the pellets in SLS. Once released, centrifugations at up to 10,300 *g* failed to sediment them.

3. Establishing the Cause of Gamma Particle Aggregation by Procedure B

Zoospores that had been collected in 1 mM Na phophate buffer at pH 7.6, concentrated by centrifugation for 3 min at 1,000 *g*, and immediately lysed by resuspension in SLS were fractionated by differential centrifugation. The organelles in such homogenates were thoroughly dispersed and showed little tendency to aggregate. Following a preliminary centrifugation for 5 min at 1,600 *g* to remove large cellular debris, sediments were obtained from 5 min centrifugations at 7,000 *g*, and especially at 9,000 *g*, which were enriched for gamma particles.

Spores that had been harvested, washed, disrupted with a glass homogenizer, diluted with water, and fractionated by differential centrifugation yielded, at 3,900 *g*, sediments enriched for gamma particles, but they were heavily contaminated with extraneous membranous material.

Differential centrifugation of zoospore lysates prepared in $MgCl_2$ produced results similar to those obtained with $CaCl_2$ present in the homogenizing solution. The gamma particles aggregated and appeared to adhere nonspecifically to cellular debris; such aggregates were pelleted out of solution by relatively short, low-speed centrifugations.

Further evidence that Ca and Mg may have caused gamma particles to adhere to cellular debris and to clump into large aggregates was obtained with spores homogenized in EDTA and differentially centrifuged. Cell particulates did not aggregate in the presence of the EDTA, nor when lysates were diluted with it. The low-speed centrifugations removed large debris and unbroken cells; the 14,000 *g* centrifugation produced a very small, tightly compacted pellet overlayed with lightly compacted membraneous material. The latter was removed by gentle aspiration with a Pasteur pipette using a Tris-buffered solution of 4.5 mM $MgCl_2$ and 5 mM $CaCl_2$ at pH 7.3. The remaining undislodged pellet could then be thoroughly resuspended in tris buffer containing 5 mM EDTA. However, when the EDTA was replaced by $MgCl_2$ and $CaCl_2$, the gamma particles immediately aggregated and adhered to cellular particulates, and therefore, the pellets could not be uniformly resuspended. At this point, addition of EDTA failed to dissociate the gamma particles from debris; however, SLS did bring about their release.

4. Isolation of Gamma Particles by Differential Centrifugation by Procedure C

Spore lysates, containing neutral red and prepared in Tris buffer with sucrose, EDTA, and KCl, but not Ca or Mg, had no visibly aggregated gamma particles, nor any that were adhering to cell debris.

The centrifugation at 750 *g* removed unbroken spores, nuclear caps, nuclei, and large debris. A small colorless pellet, produced by centrifugation at 750 *g*, contained the remaining unbroken spores and a large amount of undefinable spore debris. Gamma particles were sedimented out at 10,000 *g*. The pellet was small, red, and contained primarily gamma particles, but not all of them were brought down by this centrifugation. A small red pellet was also produced at 54,000 *g*; however, although it contained gamma particles, it was more heterogeneous than the 10,000 *g* pellet.

5. Determination of Optimum Conditions for Gamma Particle Isolation by Procedure D

It was concluded from the preceding experiments that gamma particles could be pelleted out between 10,000 and 54,000 *g*. Because of their small size, however, possible effects of the composition of isolation media and other environmental variables on their morphology had not yet been determined. Furthermore, conclusive evidence had not been obtained to show the degree to which the red-colored pellets were enriched for gamma particles.

To determine the best conditions for isolating gamma particles, zoospore suspensions were homogenized in various media, the lysates were fractionated by differential centrifugation, and pellets suspected of being enriched for gamma particles were examined by electron microscopy.

Washed 10,000 *g* pellets, derived from spore lysates prepared in medium A, contained typical gamma particles, the characteristic morphology of their electron-dense matrices being well preserved in most profiles. However, approximately 85% of the identifiable gamma particles had lost their GS membranes. The major recognizable contaminants were mitochondria. The 30,000 *g* pellet contained a greater concentration of gamma particles and was free of mitochondria, but it was contaminated with ribosomes. Here, too, approximately 85% of the gamma particle profiles lacked GS membranes.

Preparation of spore lysates in medium B, which lacked KCl as compared to medium A, significantly affected gamma particle morphology (fig. 18) as well as the relative enrichment of selected fractions (medium B, flow sheet 1, this chapter, section A, 4, a) for the organelle. Again, approximately 85% of the gamma particle profiles lacked GS membranes. But in addition, in most sections the characteristic morphology of the electron-dense gamma matrix had been changed. Instead of exhibiting its usual tightly compacted, highly electron-opaque state, it had apparently undergone degradation and become frothy. The washed 15,000 *g* pellet contained a few regions with localized concentrations of gamma particles, but overall it was not greatly enriched for them. This was contrary to expectations based on results obtained using medium A.

Profiles of gamma particles isolated in medium C closely resembled those seen in sections of whole spores (fig. 18). Unwashed 20,000 *g* pellets (medium C, flow sheet 1, this chapter, section A, 4, a) contained high concentrations of gamma particles, 75% of which still retained intact GS membranes. This may have been due, in part at least, to the higher sucrose concentration and to the omission of the washing step. The electron-dense matrix was apparently not

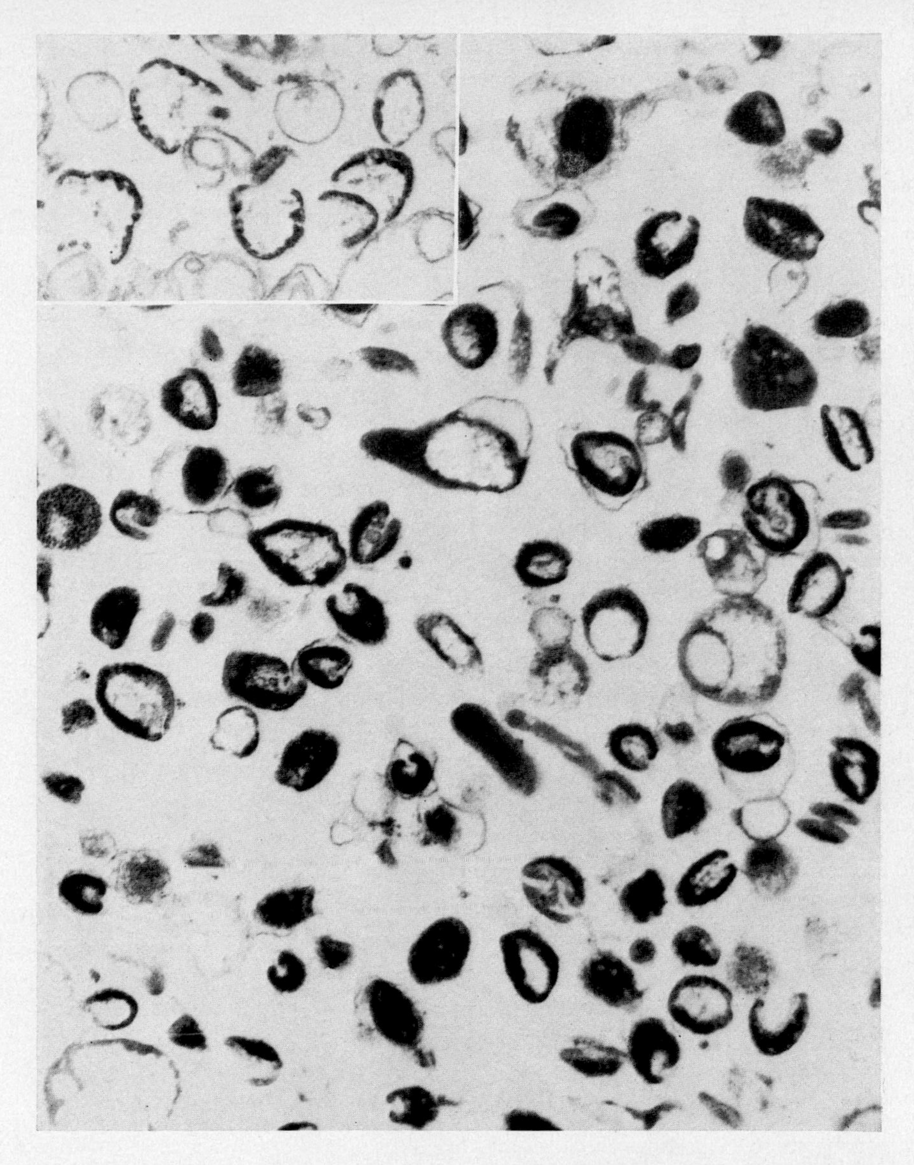

Fig. 18. Electron micrographs illustrating the appearance of gamma particles isolated in the presence and absence (insert) of KCl. X approx. 28,000 X.

altered, nor was there significant contamination by larger organelles or spore debris. The main contaminant consisted of vacuole-like bodies of gamma particle size; however, most of these probably represented sections through gamma particles which had bypassed the gamma matrix.

Solutions with the composition of medium A, but adjusted to pH. 7.4 and 8.5 (medium D and medium E, respectively), were used to test the effect of a higher pH on the isolation of gamma particles. There was no apparent alteration of the electron-dense matrices, although most profiles lacked GS membranes. The 20,000 *g* pellet (medium D, flow sheet 1, this chapter, section A, 4, a) did not contain gamma particles in very high concentrations. This contrasted sharply with the result for the corresponding fraction, prepared in medium C at pH 6.9, which was greatly enriched in gamma particles. Similarly, the concentration of gamma particles in washed 15,000 *g* fractions (medium E, flow sheet 1, this chapter, section A, 4, a) was low; they contained much vesicular debris but less amorphous material, possibly because they had been resuspended and washed. Preparations of clean and sufficiently concentrated gamma particles from these two media were not obtained.

In some of the above purification procedures, lightly packed membraneous material had been observed overlying a more compacted pellet. Had this pellet contained high concentrations of gamma particles, removal of the overlying material should have resulted in greater enrichment. However, elimination of this material did not prove worthwhile; the washed, 20,000 *g* pellets from the 10,000 *g* supernatants (medium F, flow sheet 1, this chapter, section A, 4, a) did not contain gamma particles in sufficiently high concentrations to warrant adoption of this isolation technique.

6. Final Isolation and Purification of Gamma Particles by Procedure E

As described previously (this chapter, section B, 1), the fastest and most efficient means for producing whole spore lysates, especially when large numbers of spores are required, was selective osmotic rupture of the plasma membrane. This technique minimized the amount of cellular debris that had to be separated from gamma particles during subsequent fractionation. Routine isolation and purification of gamma particles from such whole spore lysates was accomplished using a fractionation scheme slightly modified from that employing medium C (flow sheet 1; this chapter, section A, 4, a). The modification, which reduced the low-speed and high-speed centrifugation forces from 1,600 to 1,000 *g* and from 8,000 to 5,000 *g* respectively, enabled us to obtain a better recovery of gamma particles.

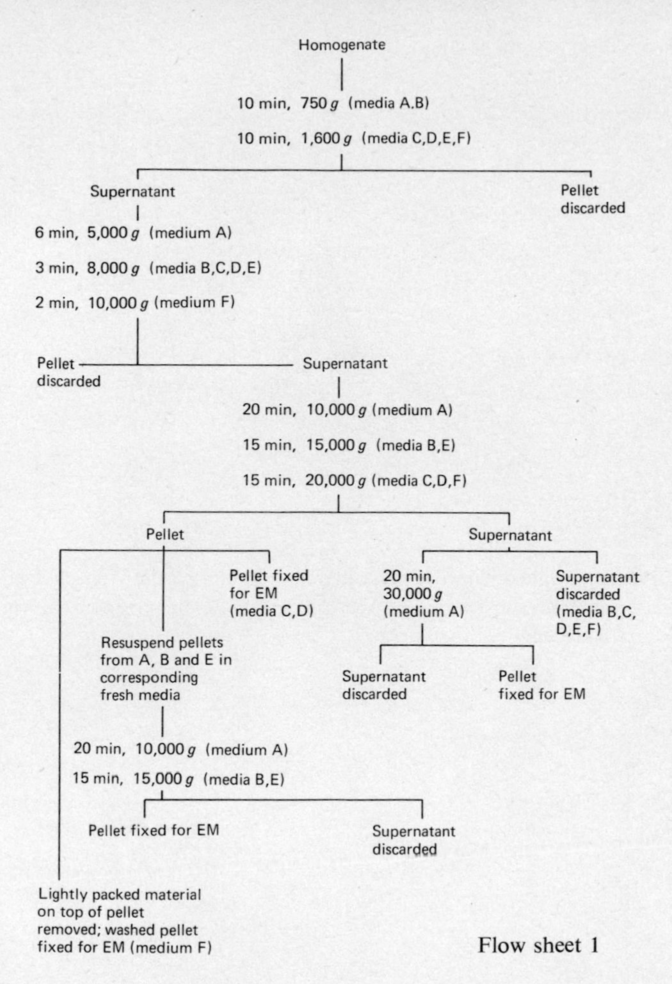

Flow sheet 1

Purification of the 20,000 *g* gamma particle enriched fractions was accomplished using sucrose density gradient centrifugation. When subjected to discontinuous gradients, gamma particles migrated through 2.02 M sucrose but layered out on top of 2.25 M sucrose (fig. 19). When isopycnic density gradient centrifugations (fig. 20) were employed, gamma particles formed a narrow, discrete band at sucrose concentrations corresponding to a buoyant density of approximately 1.32 g/cm^3. When purified by these means, gamma particles were essentially free of other cellular constituents and did not appear to be altered morphologically.

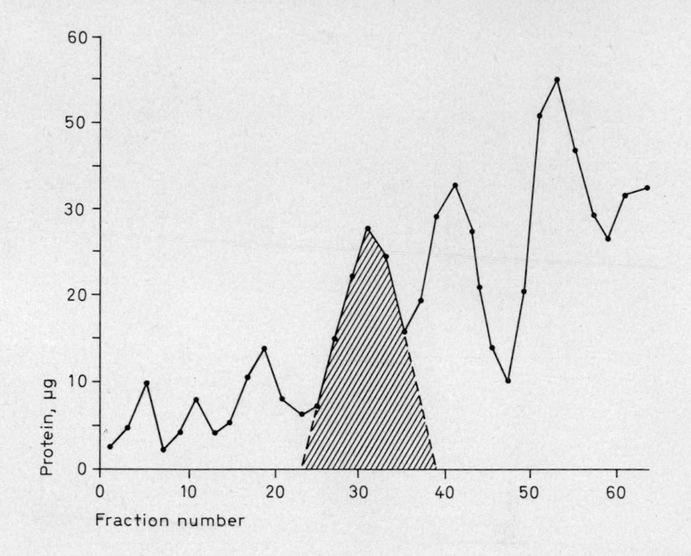

Fig. 19. Protein profile obtained after centrifugation of a partially purified gamma particle preparation through a discontinuous sucrose density gradient; region carrying gamma particles is cross-hatched.

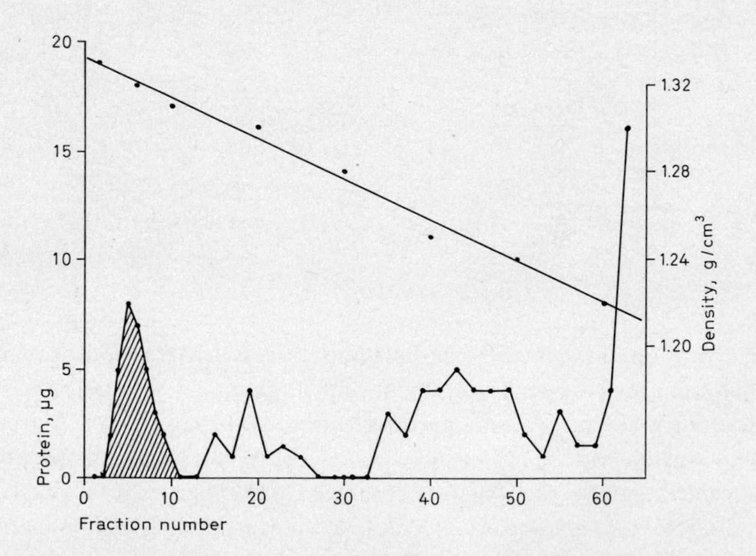

Fig. 20. Protein profile obtained after isopycnic centrifugation of a partially purified gamma particle preparation through a sucrose density gradient; region carrying gamma particles is cross-hatched.

C. Discussion

It is perhaps self-evident that several factors have to be considered before attempts are made to isolate a new subcellular organelle. Primary attention must often be given to the morphology and chemical composition of the cell from which the organelle is to be isolated. By contemplating the cell's architecture, a means for disrupting it can hopefully be devised to release the desired organelle, while minimizing contamination by other cell constituents. The isolation of gamma particles from *B. emersonii* illustrates the point. They are the only organelles that can seemingly move around freely in the cytoplasm of the zoospore; the rest are either confined by the backing membrane or anchored to other cell structures. For this reason, selective osmotic rupturing of the plasma membrane, which left the BM sheet intact, proved to be a good technique for separating gamma particles from the remaining, much larger, multi-organelle complex in the cell.

A fast, reliable means for tracking an organelle during the development of isolation procedures is also essential. For such thoroughly studied organelles as lysosomes, mitochondria, chloroplasts, nuclei, and microbodies of various sorts, this generally presents a less serious problem; in most instances, suitable markers are available such as size, morphology, enzymes, chemical composition, sedimentation characteristics, etc. For gamma particles, on the other hand, few markers had been established when our studies began. Although their approximate size had been determined, it was barely above the resolution limit of the light microscope and therefore of limited use as a marker. And even though the three-dimensional structure of the gamma particle had been established, and its electron-dense matrix shown to have a characteristic contour, routine tracking of gamma particles ultrastructurally through the various isolation procedures would have been too costly, too time consuming, and generally impractical. But as a result of having surveyed the affinity of various cytoplasmic stains for gamma particles, one of them, neutral red, proved to be useful. It facilitated the development of suitable isolation media and techniques for producing gamma particle enriched fractions.

The required degree of purity of any organelle preparation depends in large measure on its intended use. We anticipated that gamma particle preparations would eventually be used for three purposes. *Chemical characterization* would require preparations containing only gamma particles or material derived from them; they would have to be essentially free of other cytoplasmic particulates and soluble molecules, but their morphology would not have to be preserved. *Physical characterization* by way of such features as buoyant density and sedimentation velocity would not require preparations of gamma particles free of all other organelles and cell debris, but they would have to be intact morphologically and therefore in possession of their GS membrane. *Enzymic characterization*, involving elucidation of their role in the development of the fungus, would require *both* very clean and morphologically intact gamma particles.

The main criteria used thus far in assessing the degree of contamination of gamma particle preparations have been the sedimentation characteristics of the organelle and its fine structural appearance. A well-defined band of gamma particles was formed on isopycnic sucrose density gradient centrifugation. Electron-microscopic examinations in numerous fields, and at various levels when pelleted, consistently revealed high concentrations of well-preserved gamma particles. Other criteria will be employed when the organelle has been further characterized.

The effects of various factors on zoospore encystment have been studied in some detail [SOLL and SONNEBORN, 1969, 1972; TRUESDELL and CANTINO, 1971; SUBERKROPP and CANTINO, 1972]. Although direct evidence for a cause-and-effect relationship is not presently available, some preliminary correlations suggest that one such factor, namely certain cations, might alter the activities of gamma particles *in vivo*.

For example, K salts – whether they be in MOPS- [TRUESDELL and CANTINO, 1971] or Tris-maleate-buffered [SOLL and SONNEBORN, 1969, 1972] solutions – induce zoospore encystment. And, as has been emphasized (chapter IV), encystment is intrinsically associated with the decay of gamma particles *in vivo*. It may be more than mere coincidence, therefore, that the gamma matrix deteriorated *in vitro* when KCl was omitted from the medium used to isolate gamma particles, while inclusion of as little as 1 mM KCl was enough to preserve its morphological integrity.

Calcium, too, exerts effects [CANTINO *et al.*, 1968; SUBERKROPP and CANTINO, 1972; SOLL and SONNEBORN, 1972] on zoospore encystment; as seen in this chapter, Ca salts also affect the degree to which gamma particles adhere *in vitro* among themselves and to cell debris.

These observations hint at the intriguing possibility that the zoospore's hypothetical 'stabilizer' [TRUESDELL and CANTINO, 1971] or 'maintenance factor' [SOLL and SONNEBORN, 1972], its population of gamma particles, and its Ca and K components may play interrelated roles in regulating its encystment. It is a complex but unexplored area well worth pursuing further.

VIII. Chemistry of the Gamma Particle: DNA

With the development of techniques for the isolation and purification of morphologically intact gamma particles from large numbers of zoospores, chemical characterization of these organelles became possible. Their cytochemical reactions to nucleic acid stains (chapter VI) had pointed strongly to the presence of DNA. This encouraged us to look upon gamma particles as potential carriers of extranuclear genetic information. Therefore, the purpose of this investigation was to provide a more direct line of evidence for the association of DNA with the gamma particle, to develop techniques for the isolation and purification of high molecular weight DNA from the zoospores, to determine its heterogeneity and nucleotide composition, and if DNA satellites were present in the zoospore, to establish which, if any, were associated with gamma particles.[6]

A. Methods and Materials

1. Culture Methods

Massive zoospore suspensions were produced from synchronous single generation liquid cultures of OC plants using the method of SUBERKROPP and CANTINO [1973]. The spores were then concentrated at 22 °C by centrifugation at 1,000 *g* for 4 min and resuspended in a solution containing K_2HPO_4 and KH_2PO_4 25 mM (total), 1 mM KCl, and 0.25 mM EGTA at pH 6.8.

2. Assay of Gamma Particle Fractions for DNA with the Indole Method

Gamma particle preparations (20,000 *g* pellets; flow sheet 2, section VII, B, 6) were washed by resuspension in 0.3 M sucrose containing 1 mM KCl and 2 mM EDTA at pH

[6] As will be seen, certain procedures were eventually selected as the most suitable for isolating DNA from the zoospores of *B. emersonii*. However, the results of our methodological studies are presented here in some detail because – as had been the case for the isolation of gamma particles (chapter VII) – we think that they will be useful to investigators who want to do further work with the DNA in *B. emersonii*. We suggest, however, that others not so inclined proceed directly to procedure E (chapter VIII, section A, 3, e).

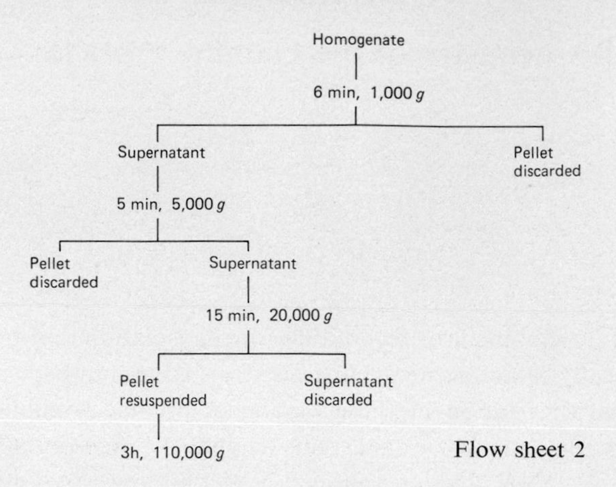

Flow sheet 2

6.5, reconcentrated at 20,000 *g* and 0–4 °C for 10 min, and assayed for DNA with indole according to SCHMID *et al.* [1963] following each of three different treatments: (1) the pellet was resuspended in H_2O, extracted 2–5 times with cold 5% trichloroacetic acid (TCA), resuspended in H_2O, and assayed for DNA; (2) the pellet was resuspended in 0.3 M sucrose containing 1 mM KCl and 2 mM EGTA at pH 6.5, incubated for 30 min at 24 °C with DNAse I (Sigma; 200 µg/ml dissolved in 2 mM Tris, 45 mM $MgCl_2$ and 5 mM $CaCl_2$ at pH 7.4) at a final concentration of 40 µg/ml gamma particle suspension, chilled at 0–4 °C, extracted five times with cold 7% TCA, resuspended in H_2O, and assayed for DNA; (3) the gamma particles in the pellet were osmotically ruptured with the procedure used by VAN BRUGGEN *et al.* [1968] for rupturing mitochondria: the pellet was resuspended in 4 M NH_4 acetate at 0–4 °C for 30 min, concentrated at 20,000 *g* for 10 min and 0–4 °C, and resuspended in H_2O. The lysed gamma particles were then digested with DNAse (70 µg/ml, extracted four times with 5% TCA, and assayed for DNA.

3. Extraction of DNA from Zoospores

Three basic procedures and two variations thereof were employed for extracting whole-spore DNA using 3×10^9–3.2×10^{10} zoospores each time. Absorption spectra and diphenylamine assays [BURTON, 1956] were used to monitor the degree of purity and concentration of the DNA preparation; the $E^{0.1\%}_{260\ nm}$ for DNA was assumed to be 20.

a) Procedure A. Either pellets of concentrated zoospores or low-speed (750 *g*) pellets of lysed zoospores were suspended in an equal volume of 100 mM NaCl containing 10 mM Tris at pH 8.0 (TS solution); this was either frozen for 1–4 days at –15 °C and then extracted, or it was extracted without freezing. In either instance, SDS was added to yield a 1% solution before extraction. After gently rolling this mixture in a round-bottom flask for 2–10 min at 24 °C, an equal volume of phenol saturated with H_2O or equilibrated with TS solution was added. The flask was rolled for another 20 min at 24 °C and the DNA-containing aqueous phase was separated by a 10 min centrifugation at 3,000 *g* and 10 °C. The phenol and denatured protein phases were suspended in 1 ml of 1% SDS or in 2 ml of TS solution, re-extracted, and recentrifuged as above. The DNA-containing aqueous phases were combined and treated in one of two ways: (a) an equal volume of H_2O- or TS-equilibrated phenol was added and the aqueous phases were then re-extracted as above, or (b) they were first clarified by a 10 min centrifugation at 300 *g* and 10 °C and then re-extracted. The aqueous phases were freed of residual phenol by shaking them with an equal volume of ethyl ether and then dialyzed against 6 liters of TS solution at 5 °C with two changes. The dialyzed preparation was hydrolyzed with pancreatic RNAse (Sigma; 1 mg/ml dissolved in 0.15 M NaCl at pH 5 and heated at 100 °C for 10 min) at a final concentration of 20–50 µg/ml for 30–60 min at 37 °C and extracted with an equal volume of phenol after adding dry SDS to yield a 0.5–1 % solution. The phases were separated by a 10 min centrifugation at 3,000–6,000 *g* and 10 °C, and the DNA-containing aqueous phase was shaken with an equal volume of ethyl ether and dialyzed against 6 liters TS solution for 16 h at 5 °C. Addition of 2 vol 95 % ethanol at 0–4 °C precipitated the nucleic acids which were collected after 20–60 min by centrifugation at 1,000 *g* for 5–15 min at 0–4 °C. The precipitate was redissolved in one tenth strength SSC (SSC is 150 mM NaCl and 15 mM Na citrate), and the DNA was selectively precipitated using MARMUR's [1961] technique by adding 0.1 vol 3 M Na acetate containing 1 mM EDTA and then slowly adding 0.54 vol isopropanol at room temperature while rapidly stirring the solution. The precipitated DNA was collected by centrifugation and then either dissolved in TS solution or first dialyzed against 6 litres of TS solution overnight at 5 °C and then dissolved. In some instances the foregoing selective precipitation of DNA was repeated.

b) Procedure B. Concentrated pellets of zoospores were lysed by gentle resuspension in an equal volume of 0.5 % SDS in TS solution (TS-SDS solution). The lysate was then either frozen at –17 °C for 1 h or combined immediately with an equal volume of phenol equilibrated with TS-SDS solution in a round-bottom flask. The flask was gently rolled for 20 min at either 5 or 22 °C. The DNA-containing aqueous phase was separated from denatured protein and phenol by two successive 10 min centrifugations, the first at 5,000 *g* and the second at 10,000 *g* for clarification, at either 4 or 22 °C. The nucleic acid was precipitated by adding 2 vol 95 % ethanol at 0–4 °C and collected by winding it onto a glass rod and by centrifuging it for 5 min at 1,000 *g* and 20 °C. It was dissolved in one tenth strength SSC and brought to full strength SSC by addition of 0.1 vol of ten times normal strength SSC. In some cases the denatured protein at the interphase was resuspended in 4 ml TS-SDS solution and re-extracted with phenol, and the nucleic acid was precipitated and dissolved as above. It was then incubated with preheated pancreatic RNAse (50 µg/ml) for 45 min at 37 °C or for 3 h at 6 °C, or it was first extracted with 1 vol ethyl ether to remove residual phenol, dialyzed against 4–6 liters SSC at 5 °C with two changes, and then incubated for 2 h at 24 °C with the RNAse (25 µg/ml). The digested mixture was then extracted with phenol, the DNA-containing aqueous phase was separated, and the DNA was precipitated with ethanol, collected by centrifugation, dissolved, brought up to full strength SSC, and dialyzed.

c) Procedure C. Concentrated pellets of zoospores were lysed by resuspension in an equal volume of 1 % SDS containing 100 mM NaCl and 100 mM Tris at pH 9. The lysate was thoroughly mixed by rolling in a round-bottom flask and then frozen. After 4 days, it was thawed and incubated with Pronase (Calbiochem; 10 mg/ml dissolved in SSC and heated at 80 °C for 20 min) for 6 h at 24 °C, 2 mg/ml being added every 2 h. The digest was dialyzed twice against 6 liters SSC, once at 24 °C for 4 h, and once at 5 °C for 15 h. The dialyzed digest was then incubated with α- and β-amylase (Sigma; 20 µg/ml of each) and preheated pancreatic RNAse (50 µg/ml) for 2 h at 24 °C. Solid SDS was added to yield a final concentration of 1% and the solution was deproteinized by adding it to an equal volume of phenol equilibrated with a modified TS solution (100 mM NaCl and 100 mM Tris at pH 9) in a round-bottom flask. The flask was rolled for 25 min at 24°C and the DNA-containing aqueous phase was separated by a 10 min centrifugation at 5,000 *g* and 24°C. The aqueous phase was dialyzed twice against 6 liters SSC, for 3 h at 24°C and for 16 h at 5°C. The DNA was then precipitated with 2 vol 95% ethanol at 0–4°C, collected by centrifugation at 10,000 *g* and 0–4°C, dissolved in one tenth strength SSC, and brought to full strength SSC.

d) Procedure D. Concentrated pellets of zoospores were thoroughly mixed with 3 vol 2% SDS dissolved in 150 mM NaCl and 100 mM EDTA and frozen at –15°C. The lysate was thawed, solid Na perchlorate was added to a final concentration of 1 M, and the solution was deproteinized by rolling it for 30 min at 24°C with an equal volume of chloroform-isoamyl alcohol (24:1). The DNA-containing aqueous phase was separated from the denatured protein and chloroform-isoamyl alcohol solution by centrifugation at 10,000 *g* and 24°C for 15 min and then incubated with preheated Pronase for 18 h at 27°C, 1 mg/ml being added three times at 2 h intervals. The nucleic acid was precipitated by addition of 2 vol 95% ethanol at 0–4 °C, collected by centrifugation at 1,000 *g*, dissolved in one tenth strength SSC and then brought to full strength SSC. The solution was incubated with preheated pancreatic RNAse (100 µg/ml) and T_1 RNAse (Sigma, 1,000 U/ml) for 90 min at 37°C and brought to 1 M NaCl with solid salt. The Pronase digestion and the deproteinization with chloroform-isoamyl alcohol were repeated. The nucleic acid was reprecipitated from the aqueous phase with 2 vol 95% ethanol at 0–4°C, redissolved in one tenth strength SSC, and freed of contaminating RNA according to MARMUR's [1961] procedure (this chapter, section A, 3, a). The DNA was dissolved in one tenth strength SSC and brought to full strength SSC. Finally, the DNA suspension was incubated for 30 min at 24 °C with α- and β-amylase (20 µg/ml of each), and then dialyzed against 6 liters SSC at 5°C for 18 h.

e) Procedure E. Concentrated pellets of zoospores were lysed by thoroughly resuspending them in 3 vol 2% SDS dissolved in 150 mM NaCl containing 100 mM EDTA. The lysate was incubated with preheated Pronase for 16 h at 27°C, 1–2 mg/ml being added 2–3 times at 2 h intervals. Solid Na perchlorate was added to a final concentration of 1 M, followed by addition of 1 vol chloroform-isoamyl alcohol (24:1), and the mixture was gently rolled at 24°C for 30 min in a round-bottom flask. The DNA-containing aqueous phase was separated by centrifugation for 15 min at 10,000 *g* and 24°C. The protein interface was suspended in 2–3 ml 2% SDS and re-extracted as above. The aqueous phases were combined, centrifuged at 10,000 *g* and 24°C, and dialyzed for 5 h at 24°C and for 17 h at 5°C against 6 liters SSC. The dialyzed solution was incubated for 2 h at 37°C with preheated pancreatic RNAse (30–50 µg/ml). After the first 1.5 h of this incubation, 6 µg each of α- and β-amylase were added. The final digest was then dialyzed against 6 liters SSC for 17 h at 5°C. The dialyzed solution was then incubated for 1 h at 37°C with T_1 RNAse (625–1,000 U/ml) and preheated pancreatic RNAse (20–50 µg/ml). Solid NaCl was added to a final concentration of 1 M, the Pronase digestion was repeated, and the mixture was deproteinized again with chloroform-isoamyl alcohol, all as described above. The aqueous phase was separated by centrifugation for 15 min at 10,000 *g* and 24°C, dialyzed twice against 6 liters SSC, once for 6 h at 24°C and then for 16 h at 5°C, and the nucleic acid was precipitated by addition of 2 vol 95% ethanol at 0–4°C. The DNA was wound on a glass rod, dissolved in one tenth strength SSC, and the solution treated in one of two ways: it was either brought directly to full strength SSC, or first freed of contaminating RNA according to MARMUR [1961] and then dissolved in the SSC.

4. DNA Melting Profiles

The thermal denaturation of DNA was carried out in Teflon-stoppered, silicone-greased, taped cuvettes. The change in absorbance was measured at 260 nm in a Beckman DU spectrophotometer. Temperature was controlled with circulating water and monitered using a microthermosensor attached to the side of the cuvette. Absorbance readings were taken at 0.5°C intervals following a 10 min temperature equilibration. Evaporation from the cuvettes was negligible and no corrections were made for water expansion.

The guanine and cytosine (G+C) content of DNA melted in one tenth strength SSC was calculated from the midpoint temperature (T_m) of the melting curve according to the formula

$$(G+C) = (T_{m\ 0.1\ SSC}/50.2) - 0.99$$

or, when higher SSC concentrations were used,

$$(G+C) = ([Tm\ SSC_x - 16.3 \log (SSC_x/SSC_{0.1})]/50.2) - 0.99$$

[MANDEL *et al.*, 1970].

Thermal absorbancy data were plotted on normal probability paper [KNITTEL *et al.*, 1968] to confirm DNA heterogeneity. Midpoints of linear segments were used to confirm the apparent DNA heterogeneity and to calculate a more precise (G+C) value than could be obtained from T_m values.

5. Caesium Chloride Equilibrium Centrifugation of DNA

CsCl equilibrium centrifugations of DNA were performed in fixed-angle rotors according to FLAMM *et al.* [1969]. Initial densities were adjusted according to the DNA component (s) being centrifuged. Preparations were centrifuged at 33,000 rpm for 65 h at 25°C in a Beckman L2-65B preparative ultracentrifuge and type 40 rotor. Fractions were collected drop-wise from the bottoms of centrifuge tubes, diluted to 0.2 ml with 10 mM Tris at pH 8.4, and their absorbancies measured at 260 nm. Buoyant densities of selected fractions were calculated according to VINOGRAD and HEARST [1962] from refractive indices measured with a Bausch and Lomb Abbe-3L Refractometer.

6. Isolation of Individual DNA Components from Density Gradients

Fractions corresponding to each DNA component were pooled and either subjected directly to CsCl equilibrium centrifugation or they were first dialyzed twice against 5 M NaCl buffered with 10 mM Tris at pH 8.0 (at 24°C for 6h; at 5°C for 16 h) and then exhaustively against one tenth strength SSC before centrifugation.

7. Caesium Chloride Equilibrium Centrifugation of Gamma Particle DNA

Gamma particle DNA was isolated and subjected to CsCl density gradient equilibrium centrifugation, both according to the method of FLAMM *et al.* [1969]. Gamma particle pellets (procedure E, chapter VII, section A, 5, b and c) were lysed by resuspension in 1% SDS buffered with 10 mM Tris at pH 8.4. To inhibit nuclease activity, and to separate protein from DNA and to denature it, solid CsCl was added to give the solution an initial density of about 1.710 g/cm³ in a volume of 4.5 ml. After removal of denatured proteins by a 10,000 *g* centrifugation for 30 min at 22 °C, the volume was corrected to 4.5 ml and the initial density was adjusted to 1.7046 g/cm³. The preparations were then subjected to buoyant density equilibrium centrifugation (this chapter, section A, 5).

B. Results

1. Association of DNA with Gamma Particles

A substance associated with gamma particle preparations purified by differential centrifugation was tentatively identified as DNA by its absorption spectrum (fig. 21) when tested with the indole reagent. Treatment of washed gamma particles with DNAse to remove any DNA that was not an integral component of these organelles eliminated two thirds of the extractable DNA (table IV); yet a significant amount – 1.8% of the total DNA in a zoospore – remained associated with the gamma particles. About 80% of this DNA could, in turn, be eliminated by first osmotically rupturing the gamma particle's surrounding membrane and then incubating the lysate with DNAse. Since the degree to which gamma particles were lysed could not be determined accurately, it is not known if the remaining 20% of the DNA was resistant to DNAse or if – as seems more likely – it represented DNA contained within unlysed gamma particles and hence inaccessible to the enzyme. In any event, the data did suggest strongly that DNA is an integral component of the gamma particle and that, when exposed, it becomes at least partially susceptible to degradation by DNAse.

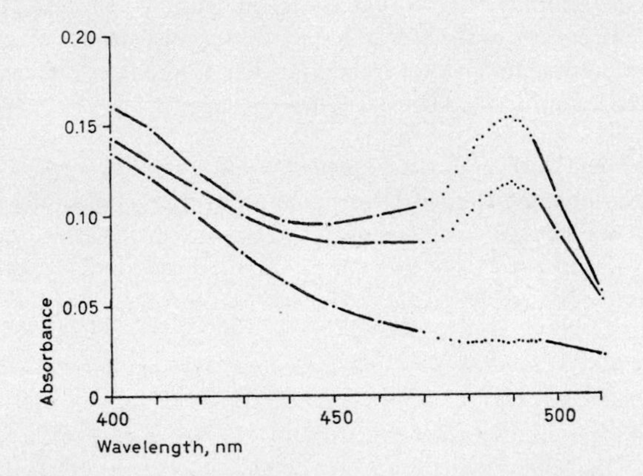

Fig. 21. Absorption spectrum (middle curve) for nucleic acid extracted from differentially purified gamma particles after treatment with the indole reagent. Top curve: deoxyadenosine standard. Bottom curve: reagent blank.

Table IV. DNA content of different gamma particle preparations as determined with the indole method

Treatment	DNA/gamma particle[1]	Gamma particle DNA/ total spore DNA, %[2]
Washed	4.3×10^{-16} g (2.6×10^{8} daltons)	6.0
Washed, DNAse-treated	1.5×10^{-16} g (0.9×10^{8} daltons)	2.1
Washed, ruptured, DNAse-treated	0.34×10^{-16} g (0.2×10^{8} daltons)	0.5

[1] Based upon a mean of 11.7 gamma particles per spore (see chapter II and footnote to table XII).

[2] Total extractable DNA per spore is about 8.3×10^{-14} g; these updated percentages are slightly different from those published earlier [Myers and Cantino, 1971].

2. Isolation of DNA from Zoospores

Because of its chemical composition, the zoospore of *B. emersonii* does not lend itself to easy isolation of DNA in an uncontaminated state. RNA and glycogen, two substances which can interfere seriously with the isolation of DNA, are also major components of the zoospore; it contains about 10–22% RNA [Lovett, 1963, 1968; Hennessy and Cantino, 1972; Suberkropp and Cantino, 1973] (table VII; chapter IX, section B, 1), the range of values being due to uncertainty about the precise weight of the zoospore, and up to 11% glycogen-like polysaccharide [Suberkropp and Cantino, 1973] but only about 0.18% DNA (table IV). For these and other reasons, we were unable to employ successfully some conventional extraction techniques used for isolating DNA in other organisms. A summary of results obtained in our attempts to isolate DNA from various kinds of starting spore material is presented in table V; they are discussed in detail in the following sections (a–e).

a) Procedure A: phenol extraction. Efficient phenol extractions of DNA from frozen zoospore suspensions were not achieved; assays of the final DNA preparations indicated that only up to 13% of the zoospore's total DNA had been recovered (table V). Elimination of RNA by selective precipi-

Table V. The results of various DNA extraction procedures

Procedure	DNA source	DNA extracted, µg[1]	260/280	Poly-saccharide[2]	Yield[3]
A	frozen zoospores	30	1.65	++++	13
	frozen 750 *g* pellet	200	2.04	++	16
	fresh[4] 750 *g* pellet	168	1.95	+	16
B	frozen zoospores	90	1.66	+++	12
	fresh zoospores	0–123	1.84–2.28	+++	0–13,88[5]
C	frozen zoospores	312	2.28	–	31
D	frozen zoospores	316	1.77	–	22
E	fresh zoospores	370–625	1.87–1.95	–	55–57

[1] Based upon the diphenylamine reaction.
[2] Relative amount of polysaccharide contamination determined by spot tests with I_2–KI.
[3] Defined as percent of total zoospore DNA.
[4] Defined as material extracted immediately without freezing.
[5] Extraction done at 22 °C, with re-extraction of first protein interface.

tation of the DNA with MARMUR's [1961] technique was effective, but after two successive isopropanol precipitations, 75% of the DNA had also been lost. The low 260/280 ratio of 1.65 suggested protein contamination. Reducing the amount of polysaccharide before phenol extraction by first lysing the zoospores to release cytoplasmic glycogen and then extracting the 750 *g* nuclei-containing pellets did not eliminate polysaccharide in the final DNA preparations, as indicated by positive I_2-KI spot tests. Freezing such 750 *g* pellets before extraction had no significant effect on the amount of DNA extracted; only 16% of the DNA was recovered by both procedures.

b) Procedure B: phenol-pH 9.0-RNAse extraction. Phenol extraction of bacterial cells in a slightly alkaline buffer has been reported [MIURA, 1967] to yield a DNA-containing aqueous phase with very little RNA contamination. A modification of this procedure was used in our attempt to isolate RNA-free DNA from zoospores. When phenol extractions were performed at 5°C, up to about 13% (table V) of the total DNA was recovered. Freezing whole spore lysates prior to extracting them with phenol did not alter the final yield of DNA. The DNA that had been extracted contained 10–20 times more RNA than DNA (as estimated from differences between $E_{260\ nm}$ and diphenylamine values), and it was also contaminated with polysaccharide. However, one variation in the procedure did increase the total DNA yield. By doing the phenol extraction at 22 instead of 5°C, the DNA recoveries approached 70%. Re-extraction of the denatured protein interface after the first phenol extraction increased the yield an additional 15%. Unfortunately, contamination by RNA and glycogen was not reduced by these modifications.

c) Procedure C: Pronase-phenol-RNAse-amylase extraction. Deproteinization of whole spore lysates with Pronase improved the efficiency of the phenol extraction. Subsequent contamination of the DNA with glycogen was prevented by using α- and β-amylase prior to the phenol extraction. Over 300 μg DNA, representing a 31% recovery of the total zoospore DNA, was achieved after a single phenol extraction (table V). However, this procedure did not eliminate contamination with RNA; therefore, the final DNA suspension was not suitable for further analysis.

d) Procedure D: chloroform-isoamyl alcohol-pronase-RNAse-amylase extraction. Deproteinization of zoospore lysates with a chloroform-isoamyl alcohol mixture, followed by a Pronase treatment of the aqueous phases, was more effective than the previous phenol-extraction procedures; up to 70% of the cell's DNA was thus recovered. Yet subsequent attempts to remove contaminating RNA reduced the yield to about 22% (table V). Furthermore, this final DNA solution was still excessively contaminated with RNA; only about 20% of its absorption at 260 nm was attributable to diphenylamine positive material. As had occurred in procedure C, the amylase treatment eliminated glycogen from the DNA preparations. The low 260/280 ratio was probably due to amylase protein which was not removed. Assays of the denatured protein resulting from the first chloroform-isoamyl alcohol extraction indicated that a significant amount of DNA had not been extracted.

e) Procedure E: Pronase-chloroform-isoamyl alcohol-RNAse-amylase extraction. Small modifications of procedure D increased the yield of DNA to over 55% (table V). Incubation of fresh zoospore lysates with Pronase *prior* to deproteinization, coupled with re-extraction of the denatured protein interface, resulted in an initial recovery of over 95% of the DNA. Inclusion of T_1 RNAse with pancreatic RNAse in two successive treatments, combined with selective isopropanol precipitation of DNA, removed most of the RNA. Over 70% of the absorption at 260 nm was due to diphenylamine positive material, and the high 260/280 ratio was indicative of complete deproteinization.

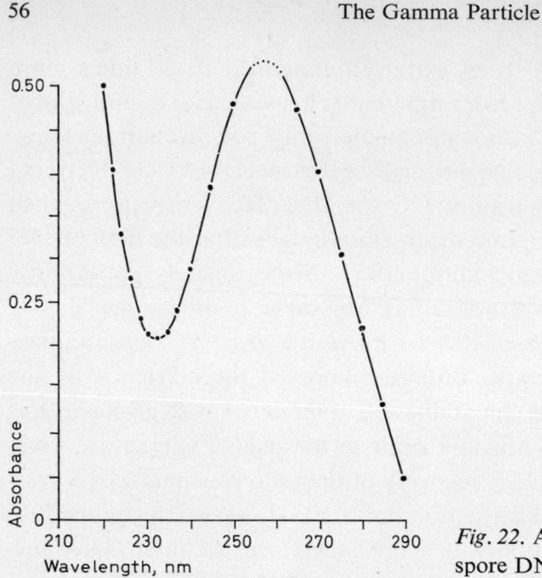

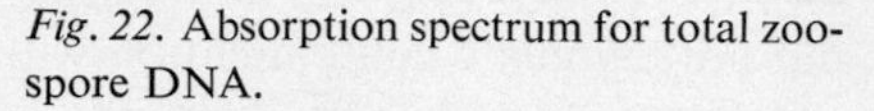
Fig. 22. Absorption spectrum for total zoospore DNA.

3. Characterization of Zoospore Nucleic Acid as DNA

The nucleic acid of the zoospore was extracted and purified by procedure E and shown to be DNA by a positive diphenylamine test and a negative orcinol reaction [SCHNEIDER, 1957], as well as detection of the four major DNA bases following acid hydrolysis [WYATT, 1951] and paper chromatography. The ultraviolet absorption spectrum (fig. 22) of a representative isolate of zoospore DNA was typical of nucleic acids with their characteristic maximum absorbancy at 257 nm. The absorbancy ratios of 260/280 nm and 260/230 nm were 1.95 and 1.97, respectively.

4. Thermal Denaturation Profiles of Zoospore DNA

Complete denaturation of whole spore DNA in full strength SSC was not possible within the temperature range (up to 94°C) employed. The DNA that did melt seemed heterogeneous and apparently consisted of more than one species. However, DNA can generally be denatured at lower temperatures by decreasing the concentration of salt in the suspending media [MARMUR and DOTY, 1962]. Thermal denaturation profiles for zoospore DNA dissolved at two lower salt concentrations are delineated in figure 23. At these concentrations, the DNA began melting at lower temperatures and the curves extended over a broader temperature range. By reducing the concentration of salt in the suspending media to one third, i.e. to 59.1 mM Na^+, the T_m was lowered and the curve was resolved into more than one step. In one tenth strength SSC, i.e. at $Na^+ = 19.5$ mm, the thermal denaturation curve exhibited four distinct steps. At this lower salt concentration the absorbancy of the total spore DNA remained constant up to 63°C. It then rose uniformly to 66°C (step I), flattened, rose again from 68 to 71.5°C (step II), rose sharply from 72 to 74°C before inflecting again (step III), and finally exhibited a steep rise (step IV), reaching its maximum at 89°C. The denaturation temperatures (T_ms) were 64.5, 69.8, 73.5 and 82°C, respectively. The denaturation profile and the hyperchromic effect (increase in absorbancy of 27%) suggested that at least the DNA corresponding to step IV was probably double-stranded. The thermal denaturation profile for *Pseudomonas putrida* DNA melted under identical conditions (fig. 23) had a T_m of 80.5°C. The thermal transition data, when plotted on probability paper, confirmed the heterogeneity of the DNA and permitted a more accurate determination of its T_m values (fig. 24).

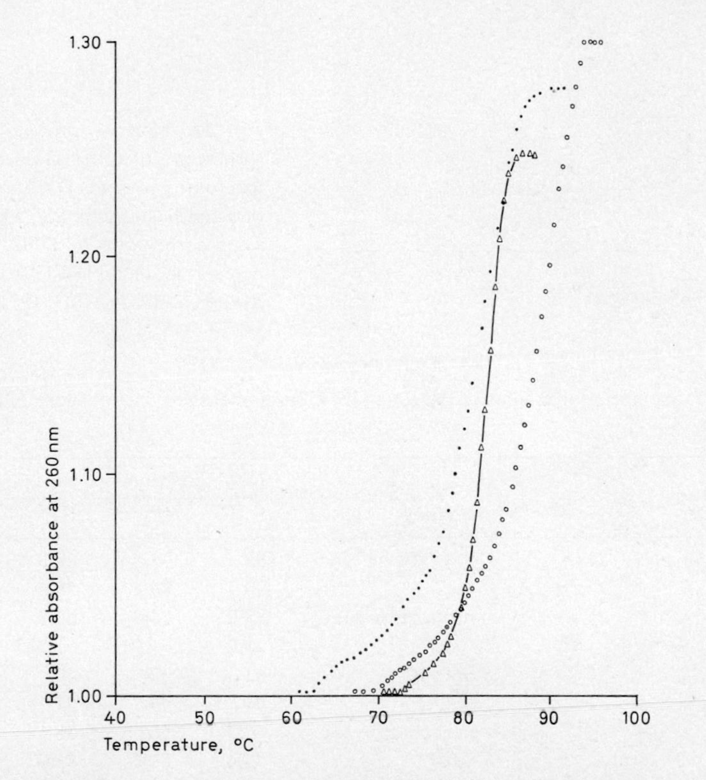

Fig. 23. Thermal denaturation of total zoospore DNA and *Pseudomonas putida* DNA. Zoospore was melted in one tenth strength (●) and one third strength (○) SSC; *P. putida* DNA was melted in one tenth strength SSC (△).

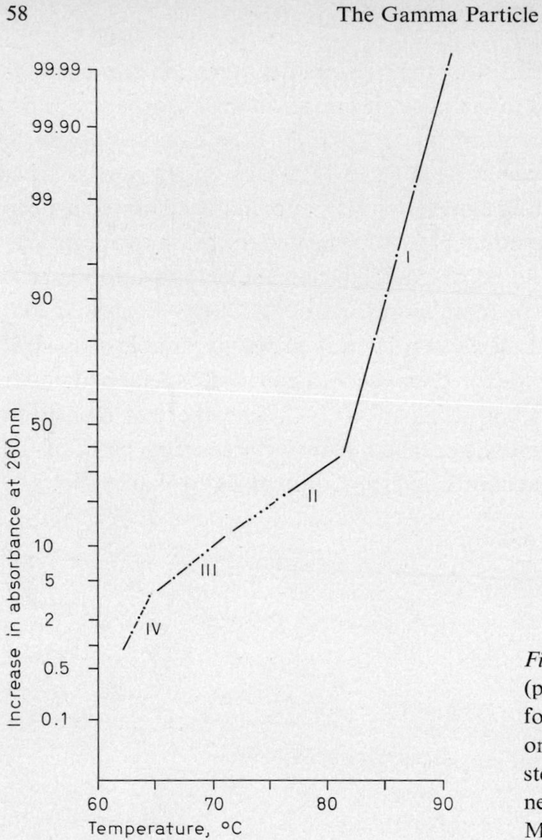

Fig. 24. Normal probability plot (procedure of KNITTEL *et al.* [1968] for total zoospore DNA melted in one tenth strength SSC. The four steps correspond to DNA components I–IV in table VI [from fig. 2, MYERS and CANTINO, 1971].

Table VI. Some characteristics of the four DNA components in the zoospore of *B. emersonii*

Parameter	Method[1]	DNA components I	II	III	IV
T_m	SSC/10	82	73.5	69.8	64.5
T_m	SSC/3.3	90	80.5	76.5	71.8
T_m	probability plot	83.2	76.8	68.8	63.5
Buoyant density	CsCl	1.731	1.715	1.705	1.687
(G+C),%	SSC/10	64	47	40	29
(G+C),%	SSC/3.3	64	43	37	27
(G+C),%	probability plot	66.7	56	38	27.5
(G+C),%	CsCl	72.4	56.2	45.0	27.6
Approximate % of total DNA	–	77.0	12.5	9.0	1.5

[1] Calculations of (G+C) levels are based on an assumed absence of odd bases.

The (G+C) contents of the four DNA components were calculated from these melting curves (table VI).

5. Buoyant Densities of Total Zoospore DNA

The buoyant densities of total zoospore DNA were determined by preparative equilibrium density-gradient centrifugation in CsCl. A typical UV absorbance profile is shown in figure 25. One major component and three lighter and distinct satellites with buoyant densities of 1.731, 1.715, 1.705, 1.687 g/cm^3, respectively, were clearly evident. The base compositions of these four DNAs, calculated from their buoyant densities [SCHILDKRAUT *et al.*, 1962], were 72.4, 56.2, 45.0 and 27.6% (G+C), respectively. The foregoing (G+C) values for components II and IV are in close agreement with those derived from corresponding T_m values (table VI). However, those for components I and III are greater than the values calculated from their T_m values. This discrepancy cannot be satisfactorily explained at present. It might be related to NASS' [1969b] observation that the formula of SCHILDKRAUT *et al.* [1962] yields values that read too high on preparative gradients were it not for the fact that, for components II and IV, the T_m and buoyant density data are almost identical. A preliminary search by paper chromatography for unusual bases in whole spore DNA hydrolysates did not resolve this question.

6. Buoyant Density Equilibrium Centrifugation of Component I

The exceptionally high buoyant density (1.731 g/cm^3) of the major DNA species was confirmed by CsCl equilibrium centrifugation of the isolated component. When DNA fractions corresponding to component I (fig. 25) were pooled and recycled in CsCl, only one peak was seen (fig. 26). The buoyant density of this DNA was 1.731 g/cm^3 and corresponded to 72.4% (G+C). As a control, *Pseudomonas putida* DNA ($\rho = 1.722$ g/cm^3), (G+C) = 63.7% [LEE and BOEZI, 1966] was similarly centrifuged to equilibrium in CsCl (fig. 26). This DNA, by our procedures, had a buoyant density of 1.724 g/cm^3, the corresponding base composition being 65.3% (G+C).

7. Thermal Denaturation Profiles of Individual DNA Components

The thermal denaturation profile for component I DNA suspended in one tenth strength SSC is given in figure 27. Its absorbance remained constant up to 74.5°C, then rose to a maximum at 86.5°C. Its T_m was 81°C. The sharpness of the helix-coil transition with increasing temperatures, and the extent of the hyperchromic shift (about 30%) indicated that it was probaly double-

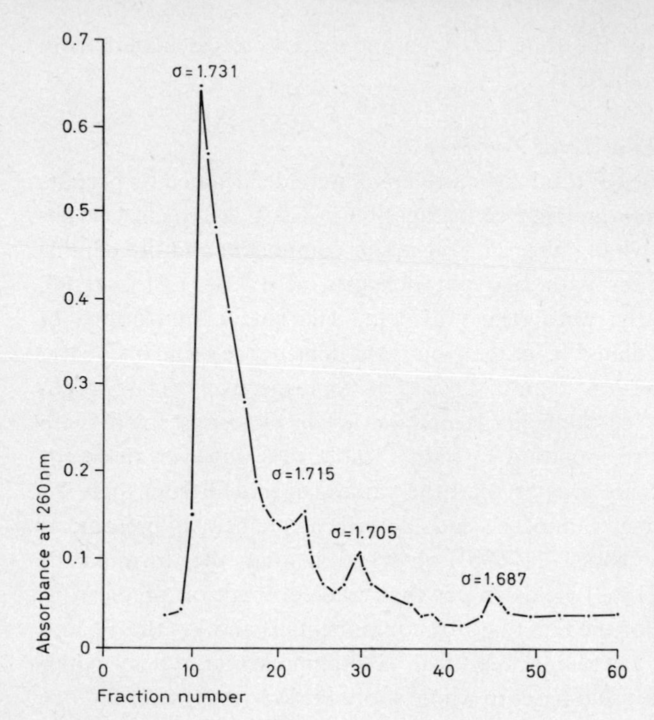

Fig. 25. Absorbance profile for 50 μg of total zoospore DNA centrifuged to equilibrium in a CsCl density gradient; the initial density was 1.7057 g/cm³.

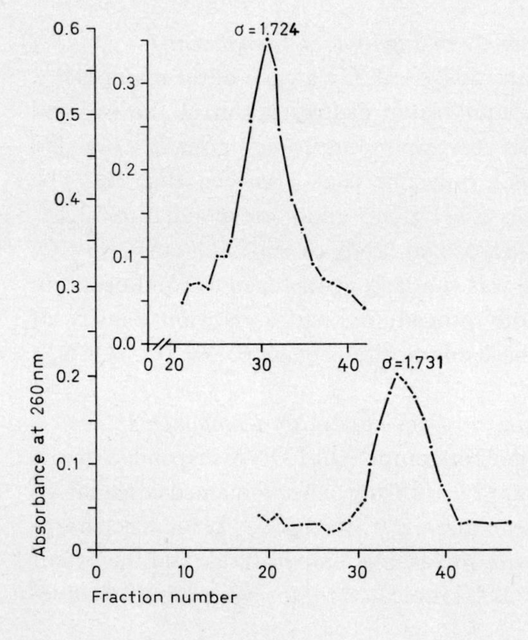

Fig. 26. Absorbance profile for 20 μg of component I DNA centrifuged to equilibrium in a CsCl density gradient; the initial density was σ = 1.7393 g/cm³. Insert: absorbance profile for 21 μg of *Pseudomonas putida* DNA centrifuged as above; the initial density was σ = 1.7100 g/cm³.

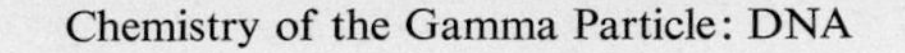

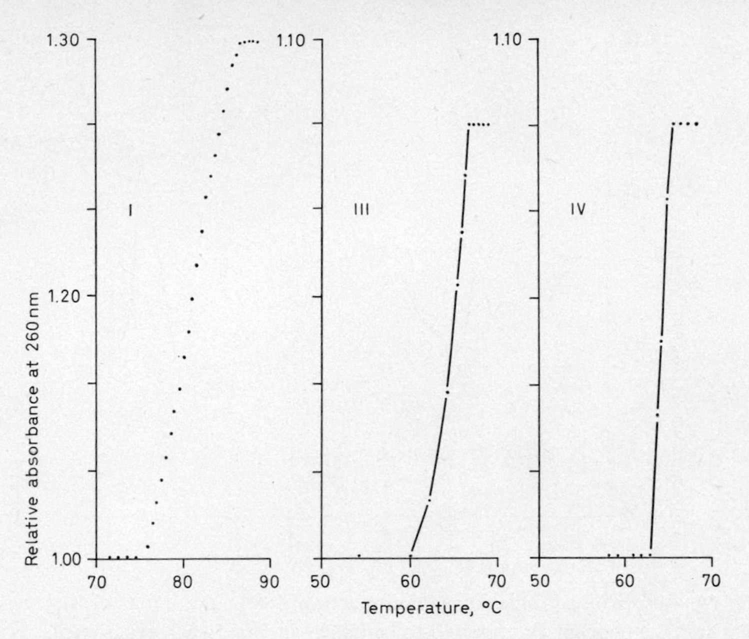

Fig. 27. Temperature-absorbance profiles for DNA components I, III, and IV denatured in one tenth strength SSC.

stranded DNA. The (G+C) of component I was calculated to be 62.3% from its T_m. Component II DNA could not be sufficiently freed of component I DNA; hence, its thermal denaturation characteristics were not determined.

However, uncontaminated samples of DNA components III and IV were isolated, and their thermal denaturation profiles were etablished (fig. 27). Their T_ms were estimated to be 66 and 64 °C, respectively. The small hyperchromic shifts exhibited by both of these DNAs in one tenth strength SSC – 12 and 10% for components III and IV, respectively – indicate that less than 25% of the DNA is double-stranded (assuming a 40% hyperchromic shift for native double-stranded DNA).

8. *Isolation and Buoyant Density Determination of Gamma Particle DNA*

Because the amount of gamma particle DNA in a zoospore is exceedingly small, conventional procedures for isolating DNA were not adequate. However, it had been shown [FLAMM *et. al.*, 1969] that some cell organelles can be simultaneously deproteinized and extracted for nucleic acids by suspending them directly in a CsCl solution. A short, low-speed centrifugation

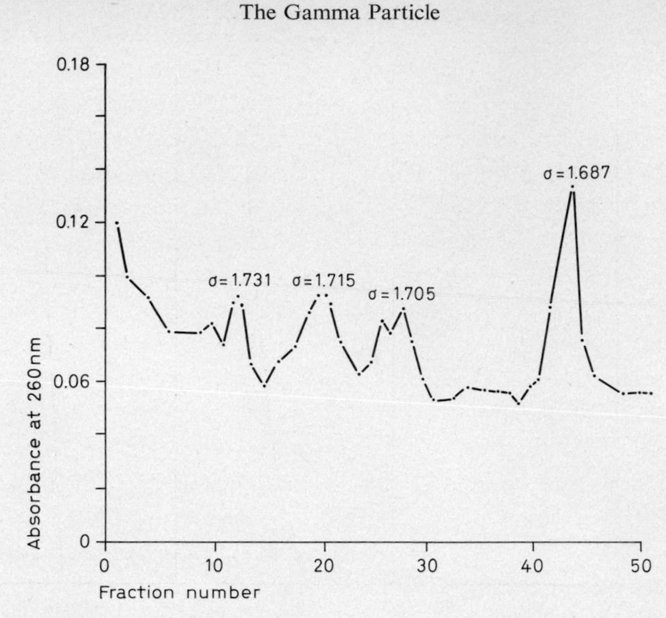

Fig. 28. Absorbance profile for gamma particle DNA; the initial density was $\sigma = 1.7057\ g/cm^3$. Component IV appeared to be binary in 2 of 5 such experiments.

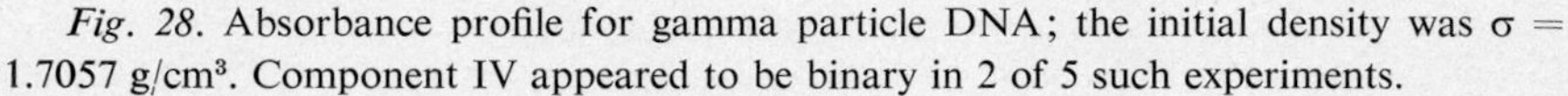

removes most of the denatured protein; then, by adjusting the volume and initial density of the CsCl solution, the organelle DNA can be isolated with negligible loss or shearing and its buoyant density can be established. Whenever we isolated gamma particle DNA by this technique, the absorbance profile after centrifugation revealed that while all three zoospore DNA satellites were present, the only one which increased was component IV; it had become enriched about 10-fold. To substantiate these results and, simultaneously, to eliminate non-DNA material, fractions corresponding to all four DNA components above were pooled and centrifuged in CsCl (fig. 28). Component IV represented about 20% of the total absorbance as compared to its earlier contribution of 1.5% (table VI); components III and II, although still detectable, were not enriched. As expected for nuclear DNA, component I was significantly reduced (cf. fig. 25).

C. Discussion

1. Association of DNA with the Gamma particle

Comb *et al.* [1964] have studied the DNA in growing plants of *B. emersonii* and concluded that it contained a major component and one detectable satellite. We have studied the DNA in its zoospores, and conclude that it is composed of a major component of very high buoyant density and three minor satellites of lower buoyant densities. Reasons have been given [MYERS and CANTINO, 1971] for believing that the major species, comprising about 77% of the total zoospore DNA, represents nuclear DNA, and that at least one of the two satellite species of intermediate density is of mitochondrial origin. The facts presented in this chapter suggest that component IV, the lightest of the three satellites and comprising about 1.5% of the total spore DNA, is associated with gamma particles. However, when washed gamma particle preparations are assayed with the indole reagent, the DNA associated with them accounts for about 5% of the total DNA in the spore. About 65% of this DNA is removed by incubation with DNAse. Whether this DNAse-susceptible DNA is simply a contaminant or actually a fraction of gamma particle DNA was not determined. It has been reported [LUCK and REICH, 1964], however, that mitochondrial membranes are impermeable to DNAse; in fact, pretreatment of mitochondrial preparations with DNAse has been used routinely in attempts to remove nuclear DNA contaminants.[7] Assuming that the GS membrane of the gamma particle is impermeable to the enzyme, the remaining DNAse-resistant DNA still accounts for about 1.8% of the total spore DNA. This figure agrees well with the 1.5% derived from the total absorbancy of component IV (fig. 25) and its proportion of the total DNA in the absorption profile resulting from CsCl density gradient centrifugation. Furthermore, when DNA is extracted from the isolated gamma particle preparations, component IV is the only one to become enriched. Since these three assays were performed independently with similarly prepared gamma particle suspensions, all the available evidence suggests that component IV is in fact DNA extracted from these organelles.

It has been said [SAGER, 1972, p. 365] that although DNAs have been reported in association with various membrane fractions from eukaryotic cells, '...as yet no rigorous evidence has identified a particular DNA with a particular membrane fraction or structure other than chloroplasts and mitochondria.' Our evidence for the reality of gamma particle DNA is still incomplete in several respects. We believe, however, that it is adequate to render its existence highly likely.

[7] But apparently not all membrane-bound, DNA-containing organelles, including mitochondria, are necessarily impermeable to this enzyme. Exogenously supplied DNAse has been shown to hydrolyze trypanosome kinetoplast DNA [LAURENT and STEINERT, 1970] and locust mitochondrial DNA [TANGUAY and CHAUDHARY, 1971].

2. Quantity of DNA in a Gamma Particle

Cytochemical detection of DNA in an organelle is dependent upon its quantity and physical state *in situ* [WOODCOCK and BOGORAD, 1970], i.e. whether it is diffuse or condensed. For a Feulgen stain to be visible by light microscopy, about 10^9 daltons of DNA are apparently required [SAGER, 1972, p. 58]. We had concluded from early results that some, but not all, of the gamma particles in a spore are visibly stained by the Feulgen dye, and that better results are obtained with azure A. Now that we know the quantity of DNA in a gamma particle, we can perhaps better explain these findings. There is about 8.3×10^{-14} g of DNA in a zoospore of which 1.5%, or 12.5×10^{-16} g, is located in the gamma particles. Since the average number of gamma particles per spore is 11.7 (chapter IV), 1.1×10^{-16} g of DNA would be contained within each organelle if it were distributed equally among them. If we assume that this DNA exists in the form of a single molecule, its molecular weight would be about 7×10^7 daltons. This is apparently below the limit of detection by the Feulgen stain, although it might be seen with the more sensitive azure A technique. Therefore, whether or not gamma particles are visibly stained *in situ* might depend upon their spacial orientation and the physical state of their DNA. Alternatively, however, the staining results might also be explained by assuming that not all of the gamma particles contain the same amount of DNA, hence some will have more, and others less, than the calculated average amount. A somewhat similar situation has been reported for *Acetabularia*, in which as little as about 20–35% of the chloroplasts contain detectable DNA [WOODCOCK and BOGOROD, 1970].

3. Potential Information Content of Gamma Particle DNA

Insofar as they contain DNA, gamma particles can now be added to the assemblage of other cytoplasmic organelles – chloroplasts and mitochondria – known to carry DNA [for compilations, see SCHIFF and EPSTEIN, 1965; COUDRAY *et al.*, 1970]. This does not mean, of course, that the mere presence of DNA in a gamma particle automatically means it must have a primary genetic function, or that it must specify structural and/or functional attributes of the gamma particle. The isolated mitochondrial DNAs of a wide variety of animal tissues consist of closed circular duplex molecules with a molecular weight of about 10^7 daltons [see SWIFT and WOLSTENHOLME, 1969 and SAGER, 1972 for recent reviews]. If gamma particle DNA is a single molecule of about 7×10^7–8×10^7 daltons, it could have nearly 10 times the information content of vertebrate mitochondrial DNA (10^7 daltons), or it might consist of highly multiple tandem repeats of shorter sequences with similar information content. Unlike mitochondrial DNA from vertebrates, the mitochondrial DNA of *Tetrahymena* and *Paramecium* [SUYAMA and MIURA, 1968], and *Physarum polycephalum* [EVANS and SUSKIND, 1971] have apparent molecular weights of about 4×10^7–6×10^7 daltons and are linear. Red bean mitochondrial DNA is also linear and has a molecular weight of about 10^8 daltons [WOLSTENHOLME and GROSS, 1968]. Even larger DNA molecules (6.2×10^9 daltons) are found in chloroplasts of *Euglena gracilis* [BRAWERMANN and EISENSTADT, 1964]. Vertebrate mitochondrial DNAs have buoyant densities of about 1.706 g/cm^3, whereas those of the larger linear mitochondrial DNAs of lower eukaryotes, such as that of *Physarum polycephalum*, have lower buoyant densities of about 1.686 g/cm^3 [HORWITZ and HOLT, 1971; EVANS and SUSKIND, 1971], and that of the chloroplast DNA of *Euglena gracilis*, 1.685 g/cm^3 [STUTZ and RAWSON, 1970]. On the foregoing bases, gamma particle DNA more closely resembles the extranuclear DNA of lower eukaryotic organisms. This amount of DNA, if it consists of unique base sequences and is transcribed and translated, should be sufficient to code for a number of proteins of average molecular weights. Our estimate for gamma particle DNA (5 µg DNA/mg protein) is higher than those reported by NASS [1969a] for differentiated tissue, but is about the same as those reported by TANGUAY and CHAUDHARY [1971] for locust muscle mitochondria (1–5 µg DNA/mg protein), and by LENNIE *et al.* [1967] for blowfly muscle mitochondria (up to 5.3 µg/mg protein).

4. Is Gamma Particle DNA Functional?

To what extent, if any, might morphological and biochemical differentiation in a zoospore be determined by a genome in the gamma particle? The loss of a cell's capacity to survive, with a concomitant loss of DNA from some organelle in that cell, is suggestive evidence that at least some of the functions of the organelle are indispensable. Certain characteristics of the zoospores derived from O thalli of *B. emersonii* suggest a possible function for gamma particle DNA. The spontaneous occurrence of O plants among populations of OC plants averages out between about 0.05 and 1%, depending upon the nature of the parental generation, its age, the medium, and other factors [CANTINO and HYATT, 1953a; CANTINO, 1969]. Zoospores derived from such O thalli have, among other things, very low (about 1%) viability and reduced numbers of gamma particles (about 8 vs. 12 for spores derived from OC thalli; fig. 4). It will be recalled (chapter II; fig. 5) that the percentage of 0 plants in a population can be greatly increased by growing *B. emersonii* in the presence of cycloheximide, and that the zoospores released from these induced ephemeral O plants also contain an average of only eight gamma particles. It had not been established whether cycloheximide might have exerted its effect directly on gamma particles, i.e., by preventing replication of their DNA, or if it acted by way of some other mechanism. However, it is known that cycloheximide specifically inhibits cytoplasmic protein synthesis [SISLER and SIEGAL, 1967; PESTKA, 1971] by preventing chain initiation and elongation; consequently, because of its dependence on the synthesis of protein on cytoplasmic ribosomes, nuclear DNA synthesis can be inhibited in intact cells of eukaryotic organisms ranging from slime molds [WERRY and WANKA, 1972] and yeasts [GROSSMAN *et al.*, 1969] to tobacco [DRLICA and KNIGHT, 1971]. When the duration of a cycloheximide treatment is sufficient to permit completion of organelle DNA synthesis by proteins made *prior* to addition of the antibiotic, the synthesis of tobacco chloroplast DNA [DRLICA and KNIGHT, 1971] and synthesis of rat liver mitochondrial DNA polymerase [CH'IH and KALF, 1969] – hence, formation of mitochondrial DNA – is also inhibited. Such findings suggest that organelle DNAs also require proteins made on cytoplasmic ribosomes for replication. Since gamma particles do not contain ribosomes, and therefore, do not possess all the essential components required for self-duplication and protein synthesis, it seems reasonable to speculate that in at least some of these organelles the synthesis of DNA was inhibited by cycloheximide before sporogenesis, the end result being fewer gamma particles, and therefore, nonviable spores.

Additional evidence that gamma particle DNA might be functional involves the activities of O zoospores and their characteristically low viability. The motile spores of *B. emersonii* do not exhibit typical gametic fusions; they do, however, form temporary cytoplasmic bridges under certain specified conditions [CANTINO and HORENSTEIN, 1954]. When 'crosses' involving such bridges were made between zoospores derived from OC plants and zoospores derived from the orange mutant BEM (chapter II), the resulting first generations of plants differed from wild-type populations in several respects [for a summary, see table IV, CANTINO and HORENSTEIN, 1954]; one of them was an increase in the viability of the mutant. These studies were made before gamma particles had been described; hence, no attempt could then be made to correlate increased viability with gamma particle counts. However, it would not be illogical to propose now, in retrospect, that the viability 'factor' being transferred might well have been gamma particles.

IX. Chemistry of the Gamma Particle: RNA

Some of the biosynthetic capacities of the zoospore of *B. emersonii* and its requirements for macromolecular synthesis during germination have been established [LOVETT, 1968; SCHMOYER and LOVETT, 1969; SOLL and SONNEBORN, 1971a, b]. Cell differentiation can proceed through but not beyond formation of the initial cyst wall in the absence of protein synthesis, whereas RNA synthesis is presumably not required until after initiation of rhizoids (however, see chapter X, section C, 4). This was interpreted [LOVETT, 1968] to mean that the zoospore contains a cryptic mRNA, the translation of which is required for differentiation beyond development of the cyst wall. And we, in turn, have provided evidence that gamma particles play an active role in the first step of the germination process by supplying at least one of the enzymes required for chitin synthesis (chapter X) and by furnishing the means for translocating this enzyme to the site of cyst wall synthesis (chapter III).

Since gamma particles contain DNA (chapter VIII) but do not carry the minimal machinery required for protein synthesis (chapter VIII, section C, 4), any message that might be produced on their DNA templates would presumably require cytoplasmic components for translation. One such component would be an RNA polymerase. Three different DNA-directed RNA polymerases have been found in zoospores of *B. emersonii* [HORGEN, 1971]. They were said to have the same general properties as three polymerases detected in OC plants [HORGEN and GRIFFIN, 1971a], one of which was considered to be a nucleolar enzyme, one was assigned to the non-nucleolar nucleoplasm, and one was apparently [HORGEN and GRIFFIN, 1971b] a mitochondrial constituent. Since zoospores contain three satellite DNAs in addition to nuclear DNA (chapter VIII), the reasonable possibility now exists that the zoospores may also possess a fourth RNA polymerase. However, no attempt has yet been made to determine if gamma particles contain a specific polymerase for transcribing gamma particle DNA, or whether genes for chitin synthetase are carried by this DNA.

With this background and the possibility that DNA-directed RNA synthesis might occur in gamma particles, we felt compelled to examine these organelles for the presence of low molecular weight RNA, especially a stable mRNA.

A. Methods and Materials

1. RNA Assay of Gamma Particles with the Orcinol Procedure

Purified gamma particles (chapter VII, section A, 5, d) were suspended in 1 ml of 25 mM K phosphate buffer containing 10 mM KCl and 1 mM EGTA at pH 6.8 and sonicated for 30 sec with 75 W at 0–4 °C. An equal volume of 15% TCA was added and the mixture was held at 0 °C for 17 h. The cold TCA-insoluble residue was collected by centrifugation at 1,000 *g* for 10 min, washed three times by resuspension in 1 ml portions of cold 5% TCA, and then extracted by heating with 1 ml 5% TCA at 95–100 °C for 20 min. RNA was determined with the orcinol procedure [SCHNEIDER, 1957] using yeast RNA as a reference standard. Cold TCA-precipitable protein was determined according to LOWRY *et al.* [1951].

2. Isolation of Gamma Particle RNA

Purified gamma particles (chapter VII, section A, 5, d) were suspended in 1 ml of 25 mM K phosphate buffer containing 10 mM KCl and 1 mM EGTA at pH 6.8 and then lysed at 24 °C by adding 1 ml of 2% recrystallized SDS dissolved in a solution of 100 mM NaCl, 1 mM $MgCl_2$ and 10 mM Na acetate at pH 6.0. The lysate was combined with an equal volume of phenol previously washed and equilibrated with the above salt solution, mixed at 5 °C for 5 min using a vibramixer, and the phases then separated by a centrifugation at 0–4 °C. The aqueous phase, which contained RNA, was decanted and re-extracted once with phenol (as above) and twice with an equal volume of chloroform-isoamyl alcohol (24:1) at 24 °C. The RNA was precipitated by addition of 2.5 vol of cold 95% ethanol and storage at –15 °C for 16 h. It was then collected by centrifugation at 10,000 *g* and 0–4 °C, dried over a stream of air, and dissolved in 100 µl of the appropriate solution. For polyacrylamide gel electrophresis, the RNA was dissolved in 5% sucrose containing 40 mM Tris, 20 mM Na acetate, 1 mM Na EDTA adjusted to pH 7.8 with glacial acetic acid, and 0.2% SDS. For RNAse treatment and nucleotide determination it was dissolved in the same buffer but without SDS. For thermal denaturation studies it was dissolved in standard SSC. The absorption spectra were determined in a Beckman DU spectrophotometer. The concentration of nucleic acid was calculated using an $E^{0.1\%}_{260\,nm}$ of 25.

3. Preparation of Polyacrylamide Gels and Electrophoresis of Nucleic Acid Species

Polyacrylamide gels were prepared essentially as described by BISHOP *et al.* [1967]. The 2.4 and 4.8% gels were prepared with Eastman chemicals, and the 7.5% gels were prepared with Bio-Rad purified chemicals. Following polymerization, the gels were removed from the Plexiglass tubing and placed in electrophoresis buffer (at about 100 ml/gel) containing 40 mM Tris, 20 mM Na acetate, 1 mM Na-EDTA, ph 7.8, and 0.2% recrystallized SDS for 72 h at 4 °C prior to use. The gels were replaced in the tubing, one end of which was covered with dialysis membrane. The gels were pre-run for 30 min at 10 V/cm and 5 mA/gel before use to remove persulfate ions from the gel interface. From 3–10 µg nucleic acid in a volume of 20–25 μl were applied to each gel and electrophoresis was carried out for either 90 min or 105 min at 10 V/cm and 5 mA/gel at 24 °C. The gel dimensions were about 0.6 × 10 cm. Immediately after electrophoresis the gels were removed from their tubes and scanned, without staining, in a Gilford gel scanner at 260 nm. *E. coli* 4S RNA was kindly provided by Dr. J. A. Boezi, Department of Biochemistry, Michigan State University.

4. Treatment of Gamma Particle RNA with RNAse

2–5 μg of gamma particle RNA in a volume of 10–25 μl were incubated with pancreatic RNAse (Sigma; 1 mg/ml dissolved in one tenth strength SSC, pH 5.0) at a concentration of 100 μg/ml for 30 min at 37 °C. The entire solution was then electrophoresed on 2.4% polyacrylamide gels.

5. Thermal Denaturation of Gamma Particle RNA

The thermal denaturation of gamma particle RNA was performed essentially as described for zoospore DNA (chapter VIII, section A, 4).

6. Determination of the Nucleotide Composition of Gamma Particle RNA

The procedures used for hydrolysis and chromatography of gamma particle RNA were essentially those described by Katz and Comb [1963]. From 1.7 to 2.1 $A_{260\,nm}$ units of RNA were hydrolyzed in 0.5 ml of 0.5 N KOH for 18 h at 37 °C. The solution was neutralized at 0 °C with 6 N $HClO_4$ and the precipitated salt was removed by centrifugation at 0–4 °C. An equal volume of 0.1 N HCl was added to the supernatant and the solution was applied to a 0.8 × 5 cm column of Dowex 50 (Hydrogen form). For elution of nucleotides, the sample was washed down with 1 ml of 0.05 N HCl and the uridylic acid (UMP) was eluted with 5 ml of 0.05 N HCl. Guanylic acid (GMP) was then eluted with 7.5 ml H_2O and cytidylic acid (CMP) and adenylic acid (AMP) were eluted together with the next 25 ml H_2O. Fractions of 0.5 ml were collected during elution of UMP and GMP, and 1.0 ml fractions were collected during elution of CMP and AMP. After bringing each fraction to 0.5 N HCl with 1 N HCl, the absorbance of fractions 1–13 (UMP) was measured at 260 nm, fractions 14–36 (GMP) at 257 nm, and the remainder (CMP+AMP) at 256 and 279 nm. The concentration of each nucleotide was determined according to Katz and Comb [1963] from its extinction coefficient in 0.05 N HCl.

B. Results

1. Total Extractable Gamma Particle RNA

A small but consistent amount of RNA is detectable in purified gamma particle suspensions when they are extracted and tested with the orcinol procedure (table VII). Such gamma particle suspensions contain about 42.7 μg RNA/mg gamma particle protein. Assuming substantial recovery (justification for this assumption is given in the footnote to table XII), of the gamma particles as well as their RNA, this would mean that there is at least about 10^{-14} g gamma particle RNA per spore, or about 0.13% of the total

Table VII. The quantities of gamma particle RNA

Method of analysis	Gamma particle RNA/zoospore, µg	Gamma particle RNA/mg of gamma particle protein, µg	Gamma particle RNA/total zoospore RNA, %
Orcinol[1]	10^{-8}	42.7[2]	0.13
$E^{0.1\%}_{260\ nm}$[3]	0.49×10^{-8}[4]	21.5[5]	0.06

[1] Analyses were done on hot TCA extracts of purified gamma particles.

[2] There is 10^{-7} µg of cold TCA-precipitable gamma particle protein per zoospore.

[3] Absorbancies were measured on phenol-chloroform-isoamyl alcohol extracts of purified gamma particles.

[4] Mean value based on analyses of gamma particle preparations derived from 20 different cultures; s. d. 0.15×10^{-8}.

[5] There is 2.3×10^{-7} µg of total gamma particle protein per zoospore.

zoospore RNA. The amount of phenol-extractable gamma particle RNA is about 50% of that obtained by the orcinol procedure; however, phenol extraction is generally considered [LOENING, 1968] to be a less efficient technique for recovering total nucleic acids.

2. Analysis of Gamma Particle RNA by Polyacrylamide Gel Electrophoresis

When electrophoresed on 2.4% polyacrylamide gels, gamma particle RNA migrated as a single component of low molecular weight (fig. 29). The major optical density peak corresponded to 4–5S RNA. Sometimes there were traces of contaminating DNA and ribosomal 25S and 18S RNA in total nucleic acid extracts of gamma particles, but most of them were essentially free of these nucleic acid species.

Electrophoresis of gamma particle RNA on 4.8% polyacrylamide gels (fig. 30) unmasked an additional optical density peak. Most of the RNA still migrated as an apparently homogeneous species, but a second, 'minor' component of lower molecular weight was detected. This component migrated slightly ahead of the major component; we have not excluded the possibility that it represents a breakdown product.

The two gamma particle RNA components were separated even better in 7.5% polyacrylamide gels (fig. 31). Since gels of this concentration have been used to separate 4S RNA from 5S RNA [ZEHAVI-WILLNER and DANON, 1972], it seemed likely that a more accurate estimate of the size of gamma particle RNA could be obtained by coelectrophoresing it with *E. coli* 4S

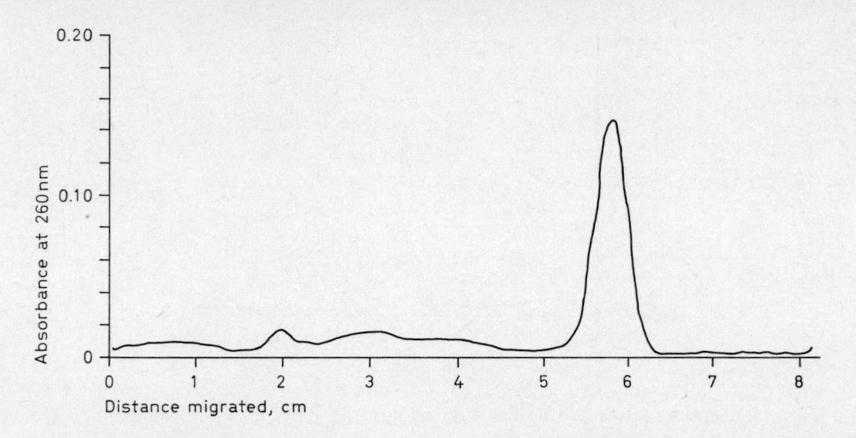

Fig. 29. Representative profile for 5 μg of gamma particle RNA electrophoresed in 2.4% polyacrylamide gel for 90 min.

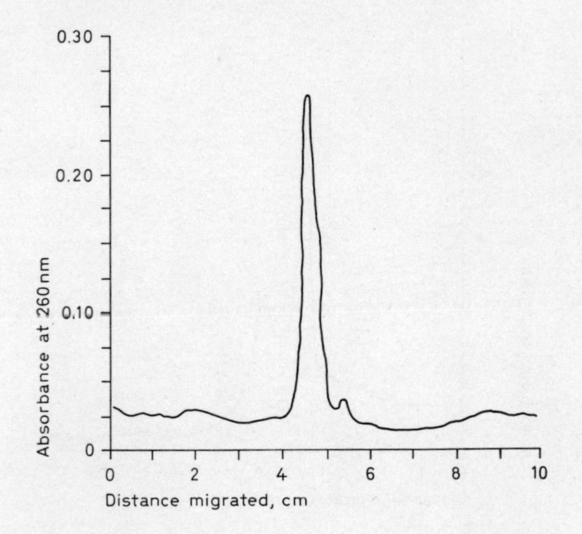

Fig. 30. Representative profile for 6 μg of gamma particle RNA electrophoresed in 4.8% polyacrylamide gel for 90 min.

RNA. The 4S RNA marker comigrated with the major gamma particle RNA component (fig. 32); hence, most of the gamma particle RNA appeared to be a 4S species, but a 'minor' component of lower molecular weight was also present. It seems unlikely that gamma particle RNA is a breakdown product of rRNA for several reasons: the presence of 2% SDS should have inhibited

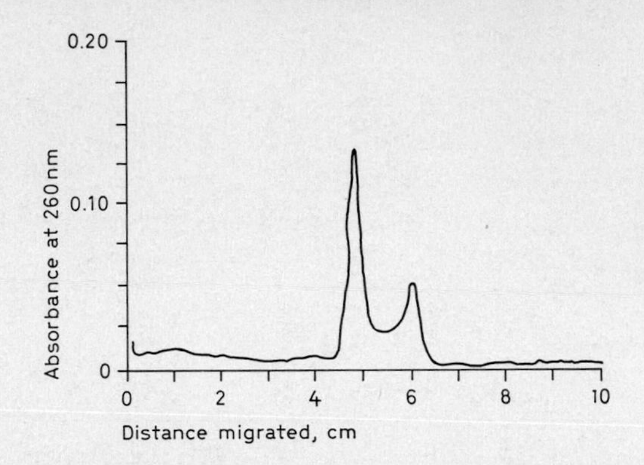

Fig. 31. Representative profile for 4 µg of gamma particle RNA electrophoresed in 7.5% polyacrylamide gel for 105 min.

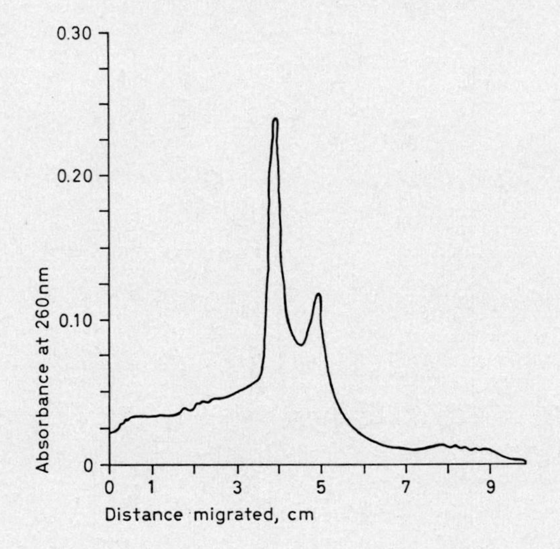

Fig. 32. Representative profile for 4 µg of gamma particle RNA and 3.5 µg of *E. coli* 4S RNA electrophoresed simultaneously in 7.5% polyacrylamide gel for 105 min.

ribonuclease activity; the amount of RNA recovered was proportional to the number of gamma particles extracted; and the RNA always migrated as a sharp peak and at a rate equivalent to that of 4S RNA. Because we were severely limited by the small quantities of gamma particle RNA that could be isolated, this 'minor' component was not further characterized.

3. Effect of RNAse on Gamma Particle RNA

Gamma particle RNA, preincubated with pancreatic RNAse, was electrophoresed on 2.4% polyacrylamide gels (fig. 33). The RNA appeared to be very resistant to enzymic attack since its migration profile differed only slightly from that of the untreated RNA. The former RNA migrated a little more slowly, its absorption profile was broader, and there was an increase of about 50% in total UV absorbance (hyperchromicity). Whenever the small amounts of contaminating ribosomal 25S and 18S RNA were present, they were subsequently completely eliminated by pancreatic RNAse, thus indicating that the enzyme was active. Therefore, although the rRNA degradation products might have contributed slightly to the broadened optical density peak in some instances, this could not have been so in all of them.

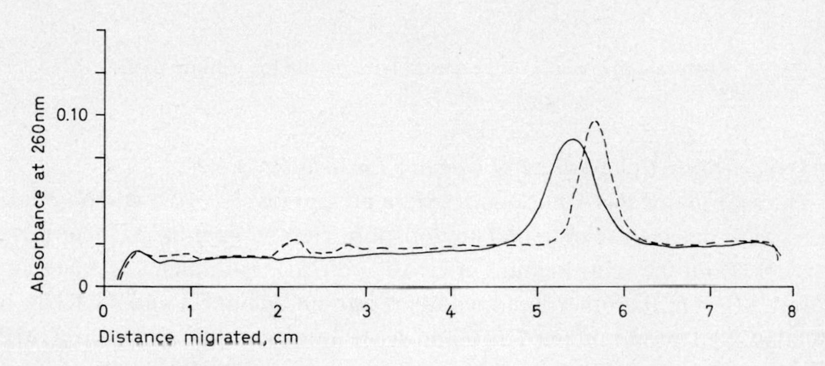

Fig. 33. Representative profiles for 3 μg of gamma particle RNA with (——) and without (– – – –) pretreatment via RNAse and electrophoresed separately in 2.4% polyacrylamide gel for 90 min. In four separate experiments, the distance between the two peaks was 0.32, 0.64, 1.27 and 1.59 cm.

4. Thermal Denaturation of Gamma Particle RNA

To find out if gamma particle RNA had secondary and tertiary structure, it was heated in a salt solution of high concentration. Its thermal transition curve (fig. 34), obtained by following the absorbance of its major bases at 260 nm in the presence of 165 mM Na^+, was clearly multiphasic; this is typical [Cole *et al.*, 1972] of transfer RNA species melted at this particular salt concentration. At least three thermal transitions (hyperchromic shifts) were resolvable, with approximate T_m's of 31.5, 40 and 52.5 °C. Gamma particle RNA has a 25% hyperchromicity.

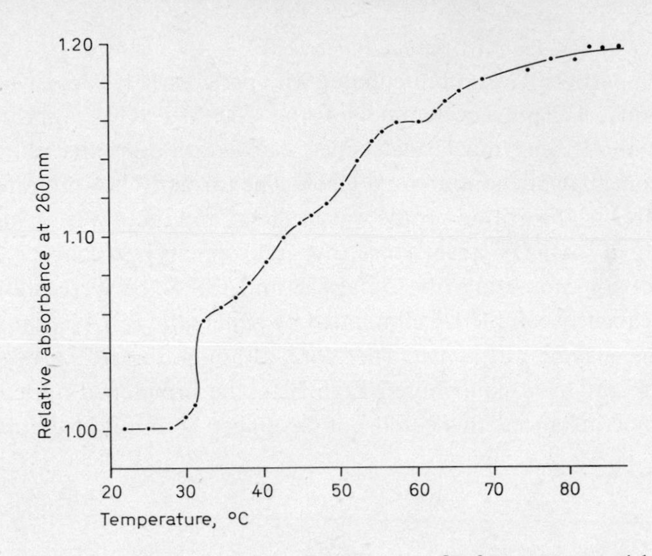

Fig. 34. Representative temperature absorbance profile for gamma particle RNA.

5. Nucleotide Composition of Gamma Particle RNA

The four major RNA nucleotides were present (table VIII), and no minor bases were detected. Compared to zoospore rRNA, gamma particle RNA contains about the same amount of GMP and CMP but much less AMP and more UMP. On the other hand, compared to OC-plant 5S and 4S RNA, it contains similar amounts of GMP and AMP but less CMP and more UMP. Clearly, the nucleotide composition of gamma particle RNA differs from that of both the rRNA in zoospores and the soluble RNAs in OC plants.

C. Discussion

1. Gamma Particle RNA: A Distinct Population of Transfer RNA Species?

In many organisms, mitochondrial 4S RNAs differ from their cytoplasmic counterparts in nucleotide composition [Nass and Buck, 1970], degree of methylation, and the absence of minor bases [Dubin and Brown, 1967; Dubin and Montenecourt, 1970]. The data available at this time permit a comparison of gamma particle 4S RNA with OC plant 4S RNAs only on the basis of their thermal denaturation characteristics and their nucleotide composition. Its lower T_m values (fig. 34) indicate that it may have a less stable secondary structure than the 4S RNA in OC plants. This, together with its

Table VIII. The nucleotide composition of gamma particle RNA and other RNAs

Nucleotide	Zoospore rRNA[1]	OC plant 5S RNA[2]	OC plant 4S RNA[3]	Gamma particle 4S RNA[4]
GMP	28.1	32.5	32.9	30.6
AMP	27.3	19.2	16.0	17.8
CMP	19.4	24.6	25.5	19.1
UMP	25.2	23.7	25.5	32.6

[1] LOVETT [1963].
[2] COMB and ZEHAVI-WILLNER [1967].
[3] COMB and KATZ [1964].
[4] Mean values for three experiments. The standard deviations for GMP, AMP,CMP and UMP are 2.74, 0.98, 0.47, and 2.66, respectively.

higher UMP and lower CMP contents (table VIII), makes it likely that the zoospore contains a distinct population of 4S RNAs, and that this particular group of 4S RNAs can be assigned unambiguously to a well-defined structural entity in the organism, i.e. the gamma particle.

2. Relation Between the DNA and RNA Contents of a Gamma Particle

DNA-RNA hybridization techniques have been used to study the degree of homology between certain DNA and RNA molecules. They have been particularly helpful in identifying transcription products and in establishing the coding potential of organelle DNAs. Homology has been demonstrated between mitochondrial DNA and mitochondrial 4S RNA in animal cells [NASS and BUCK, 1970; DAWID, 1970; ALONI and ATTARDI, 1971a, b; DAWID and CHASE, 1972], including *Tetrahymena* [CHI and SUYAMA, 1970], between chloroplast DNA and chloroplast 4S RNA in algal cells [SCHWEIGER, 1970], etc. In general, mitochondrial DNA has about 12–15 cistrons for tRNA species and chloroplast DNA has about 20 cistrons for tRNA. Because of the large portion of the mitochondrial genome that is homologous with mitochondrial RNA, it has been suggested [DAWID, 1972; ALONI and ATTARDI, 1971b] that this might be the sole function of animal cell mitochondrial DNA.

Gamma particle 4S RNA has a G+C content of 49.7%, which differs sharply from that of gamma particle DNA (28%; chapter VI). However, if this 4S RNA is transcribed from only a small percent of one strand of the entire DNA genome (homologies from $<$1–15% have been reported [CHI and SUYAMA, 1970; DAWID and CHASE, 1972] for mitochondrial DNA), its G+C level will not necessarily reflect that of the entire G+C content of the DNA molecule. The nucleotide composition of gamma particle 4S RNA is assymmetric in that the A:U ratio is about 1.8. This fact is consistent with the observation that gamma particle DNA appears to be only partially double-stranded (chapter VIII, section B, 7). We have previously made the analogy that the size of gamma particle DNA more closely resembles that of extranuclear DNAs of lower eukaryotic organisms than that of the mitochondrial DNA of higher organisms (chapter VIII, section C, 3). By extending this parallel one step further, it follows that – if ribosomes were to become available to them – gamma particles would have the potential to play a role in protein synthesis.

3. Possible Significance of a Pool of tRNA in Gamma Particles

The rate at which specific proteins can be synthesized is dependent, in part at least, upon the availability of tRNA species. Should a cell require rapid synthesis of one or a few specific proteins, it is logical to suppose that it would have readily available a localized 'pool' of the required tRNA species. A somewhat similar condition has been shown to occur in cotton seed embryos where there is a substantial increase in the amount of chloroplast tRNA during germination [MERRICK and DURE, 1972]. The non-decaying gamma particle in the zoospore of *B. emersonii* possesses DNA and tRNA, but it has no detectable ribosomes. There is at present no reason, therefore, to assume that it can itself synthesize protein. Before a zoospore germinates, the compartmentalization of certain tRNAs within the gamma particles may serve to prevent translation by spatially sequestering such RNAs from the ribosomes in the cell. Subsequently, during zoospore germination, rapid synthesis of protein does occur [LOVETT, 1968; SOLL and SONNEBORN, 1971a]. Whether such synthesis actually involves gamma particles has not been determined; but, in view of their established role (chapter III) in encystment, it seems to us that this is now a distinct possibility.

X. Enzymology of the Gamma Particle

Although gamma particles resemble lysosomes [DE DUVE *et al.*, 1955] in the general sense that both are apparently delimited by only one unit membrane (however, see fig. 9 legend) and both are about the same size, they are decidedly different in more definitive aspects of their structure and function (see discussion in this chapter, section C, 2). In view of the well-known association of acid phosphatase activity with lysosomes [BARRETT 1972], we wanted to find out whether or not this enzyme was also associated with gamma particles; its absence would have constituted yet another difference between these two structures.

A second category of organelles bound by a single membrane is the microbodies [HRUBAN and RECHCIGL, 1969], including those varieties that qualify biochemically as 'glyoxysomes' and 'peroxisomes' (see discussion in this chapter, section C, 2). One enzyme almost always associated with microbodies is catalase. Reasoning as before (that is, should the gamma particle be thought of as a sort of microbody?), we also tested gamma particles for the presence of catalase activity.

Finally, in view of the established link (chapter III) between decay of gamma particles and synthesis of wall material during zoospore encystment, as well as its logical ramification [TRUESDELL and CANTINO, 1971], gamma particles were also examined for the presence of some of the enzymes expected to be involved in the synthesis of chitin, especially chitin synthetase. At this phase of our studies, the intent was not to make an exhaustive examination of kinetic and other properties of the chitin synthetase in *B. emersonii*, for extensive work had already been done with it in SONNEBORN's laboratory [CAMARGO *et al.*, 1967]; rather, our purpose was to provide reasonable evidence that the enzyme activity which we would be measuring in gamma particle fractions was, in fact, due to chitin synthetase.

A. Methods and Materials

1. Assays for Catalase Acitivity

Gamma particles were assayed for catalase activity using two procedures, authentic catalase (H_2O_2:H_2O_2 oxidoreductase, EC 1.11.1.6; Boehringer) being used as the control. Method 1 [LÜCK, 1963]: the 1 ml reaction mixture contained 25.8 mM KH_2PO_4 and 40.7 mM Na_2HPO_4 adjusted to pH 7, 12.5 mM H_2O_2 and either (a) 8.5–25.5μg gamma particle protein – the gamma particles having been prepared by differential centrifugation of zoospore lysates (chapter VII, sections A, 5, a–c) and stored at –17 °C for several days prior to the assay –, or (b) 1.2–13.2 μg gamma particle protein – the gamma particles having been purified by isopycnic sucrose density gradient centrifugation (chapter VII, section A, 5, d) and assayed the same day. The reaction was monitored spectrophotometrically at 240 nm and 23 °C. Method 2 [QUAIL and SCANDALIOS 1971]: activity was followed via O_2 evolution with a Clark oxygen electrode at 23 °C in a 5 ml reaction mixture containing 22 mM H_2O_2 and 12 μg gamma particle protein prepared as in (b) above.

2. Cytochemical Assays for Acid Phosphatase Activity

Two cytochemical procedures were used for acid phosphatase (EC 3.1.3.2). Method of Gomori [MCMANUS and MOWRY 1960]: PYG agar cultures were flooded with H_2O and the zoospores were concentrated by centrifugation. They were fixed in 4% formaldehyde containing 90 mM $CaCl_2$ for 24 h at 0–5 °C, transferred to 0.88 M sucrose, left at 0–5 °C for 96 h, concentrated by centrifugation, resuspended in 0.25 M sucrose, air dried on microscope slides, and then incubated with the appropriate Na glycerophosphate and 'yellow ammonium sulfide' solutions. Control slides lacking substrate were run simultaneously. Method of Grogg and Pearse [MCMANUS and MOWRY, 1960]: spores prepared as above were fixed in 0.1 M Na cacodylate-HCl buffer at pH 7.2 containing 2% glutaraldehyde, pelleted, resuspended in an acid phosphatase activator buffer (0.1 M citrate, pH 3) at 0–4 °C for 30 min, recentrifuged, suspended in H_2O, air dried on microscope slides, and then incubated with the appropriate α-naphthyl phosphate and diazo fast blue B solutions. Spores were also tested for acid phosphatase activity after up to six successive Freezings and thawings. Control slides lacking substrate were also examined.

3. Assays for Acid and Alkaline Phosphatase Activity in Isolated Gamma Particles

The modified method of KIND and KING [1954], as described in the 'Phosphazyme' assay by Harleco Co., Philadelphia (literature No. 2150; Dec. 1970), was used for both acid and alkaline (EC 3.1.3.1) phosphatases. Gamma particles were isolated and purified with standard procedures (chapter VII, sections A, 5, a–d), reconcentrated by centrifugation at 110,000 *g* for 15 min at 0–4 °C, resuspended in 2 ml 67 mM K phosphate at pH 7, and assayed for phosphatases.

4. Tests for Chitin Synthetase Activity

The assays for this chitin UDP acetylaminodeoxyglucosyl transferase (EC 2.4.1.16) were made by measurement of either uridine-5′-diphosphate (UDP) liberated [PORTER and JAWORSKI, 1966] from UDP-N-acetyl-D-glucosamine (UDP-GLcNAc) during synthesis of chitin, or radioactivity incorporated into chromatographically immobile material from UDP-[$^{14}C_1$]-GlcNAc.

When activity was to be determined via release of UDP, 1 ml reaction mixtures containing 2.5 mM UDP-GlcNAc, 1.1 mM GlcNAc, 4 mM Tris at pH 7.5, 0.3 mM EDTA, 3 mM mercaptoethanol, 1 mM $MgCl_2$, and 5–900 μg cell extract protein were incubated at 30 °C for periods up to 60 min. Reactions were terminated by immersing assay tubes in boiling H_2O for 3 min after which they were cooled in an ice bath. The UDP liberated was measured as follows: 50 µl of 200 mM KCl containing 5 mM phosphoenolpyruvate and 0.1 ml of ATP: pyruvate phosphotransferase, EC 2.7.1.40 (crystalline pyruvate kinase, Type II, 4 mg/ml; Sigma) diluted 1:100 in 0.1 M $MgSO_4$ were added and the tubes were incubated at 37 °C for 15 min; 0.15 ml of 5 mM 2,4-dinitrophenylhydrazine in 2 N HCl was added; after 5–10 min incubation at room temperature, 3.6 ml of 0.556 M NaOH in 90% ethanol were added; finally, reaction mixtures were clarified by centrifugation and their absorbancies were determined at 520 nm. UDP standards and appropriate blanks were treated similarly.

When activity was to be determined via the amount of radioactivity incorporated into chromatographically immobile product, 50 µl reaction mixtures containing 1 nmol of UDP-[$^{14}C_1$]-GlcNAc (specific activity = 42 mC/mmol), 2.53 mM UDP-GlcNAc, 22 mM GlcNAc, 40 mM Tris at pH 7.5, 10 mM EDTA, 30 mM mercaptoethanol, and 4–53 µg of cell extract protein were incubated at 25 °C for periods up to 60 min. Reactions were terminated by adding 5 µl of glacial acetic acid. The entire incubation mixture was chromatographed (descending) on Whatman No. 1 paper, either for 24 h using isoamyl alcohol-pyridine-H_2O (1:1:0.8), or for 16 h using ethanol and 1 M NH_4 acetate (7:3), pH 3.8. The distribution of radioactivity on the paper was determined with a Tracerlab 4-Pi scanner; the radioactivity at the origin was also estimated by direct counting with a Tracerlab Versa/Matic II scaler and FD1-PI flow counter. Protein was determined by the method of LOWRY *et al.* [1951].

5. Test for Reutilization of UDP During Chitin Synthesis

Zoospore homogenates (100–225 µg protein) and 20,000 *g* gamma particle enriched pellets (25–60 µg protein) were incubated in reaction mixtures containing 1.1 mM GlcNAc, 4 mM Tris at pH 7.5, 0.3 mM EDTA, 3 mM mercaptoethanol, 1 mM $MgCl_2$, and 55 nmol UDP. After different time periods, reaction vessels were put in boiling water for 3 min and the contents were then assayed for residual UDP.

B. Results

1. Absence of Catalase Activity in Gamma Particles

Assays for catalase activity were consistently negative; there was no apparent activity in gamma particles prepared either by differential centrifugation (method 1, a) or by isopycnic sucrose density gradient centrifugation (method 1, b). Analyses with the more sensitive oxygen electrode procedure (method 2) also failed to demonstrate catalase activity in purified gamma particle suspensions. Control assays with authentic catalase were always positive.

2. Absence of Phosphatase Activity in Gamma Particles

Gamma particles did not react visibly to either the Gomori or the Grogg and Pearse cytochemical procedures. Repeated freezing and thawing of the spores in the citrate buffer failed to activate any latent phosphatases. On the other hand, the lipid globules (fig. 6, 11) in the SB-lipid complex did react positively [CANTINO and MACK, 1969, fig. 6] to both cytochemical procedures, thus providing a control for the tests and some assurance that the negative reaction of the gamma particles was real. The negative results were confirmed when partially purified gamma particles were assayed for acid and alkaline phosphatase activities; again, they apparently lacked both of these enzymes.

3. Presence of Chitin Synthetase Activity in Zoospores: Identification of the Reaction Product as Chitin

Both whole spore homogenates and purified gamma particle preparations were used as sources of chitin synthetase for the characterization of the immobile radioactive products formed by incorporation of label from UDP-[$^{14}C_1$]-GlcNAc. After chromatograms had been scanned for radioactivity, the labeled areas at the origin were cut out and tested as follows.

a) Test 1. The excised paper was macerated in 2 ml of 50 mM K phosphate adjusted with acetic acid to pH 5.3 and containing 10 mg chitinase (Calbiochem), and then incubated at 37°C for periods up to 28 h. The reaction was terminated by transferring 0.5 ml of the above assay mixture to 2 ml ethanol, heating it to 80°C for 2 min, and centrifuging it. The ethanolic solution was concentrated under an air stream and chromatographed (descending) on Whatman No. 1 paper using 1-butanol:pyridine:H_2O (6:4:3). The major detectable radioactive products of the chitinase hydrolysis migrated like diacetylchitobiose and acetylglucosamine. These two compounds accounted for about 30% of the ^{14}C at the origin.

b) Test 2. The excised paper was macerated in 3 ml of 25 mM K phosphate containing 1 mM KCl and 0.5 mM EGTA at pH 6.8 and incubated at 30°C for 1 h with 20 µg each of α- and β-amylase. The assay mixture was centrifuged and the supernatant was concentrated under an infrared lamp and then chromatographed (descending) on Whatman No. 1 paper using 1-butanol:pyridine:H_2O (6:4:3). No mobile radioactive substances were detected. Since the radioactive material at the origin was not significantly hydrolyzed by the two amylases, $\alpha(1\rightarrow4)$-glycans, such as the glycogen-like polysaccharide [CANTINO and GOLDSTEIN, 1961; LESSIE and LOVETT, 1968; CAMARGO *et al.*, 1967] known to be present in *B. emersonii* zoospores, had not become labeled.

c) Test 3. The excised paper was hydrolyzed in 3 ml of 6 N HCl at 100°C for 4 h. The hydrolysate was clarified by centrifugation. At least 75% of the initial radioactivity was recovered in solution. The HCl was evaporated at 24°C under vacuum; the residue was dissolved in H_2O, reconcentrated under an infrared lamp, and chromatographed as in test 1. The radioactive hydrolysis product migrated as a single component with an rf corresponding to that of glucosamine.

4. Release of UDP from UDP-GlcNAc During Chitin Synthesis

Tests for chitin synthetase activity were made with both sonicated zoospore lysates and 20,000 *g* pellets prepared by differential centrifugation (as in chapter VII, sections A, 5, a–c) of zoospore homogenates. Both preparations released UDP from UDP-GlcNAc during incorporation of GlcNAc into chitin (fig. 35). The whole zoospore lysate did so at a substantial rate. The 20,000 *g* pellet, which contained a high concentration of gamma particles, released UDP at a much higher initial rate and showed no tendency to plateau during the 45 min assay period. These specific activities are of the same order of magnitude as those obtained by PORTER and JAWORSKI [1966] for UDP released from UDP-GlcNAc by 'microsomal' and 'mitochondrial' fractions from *Allomyces*, a relative of *Blastocladiella*.

5. Comparative Chitin Synthesis by Different Cell Fractions as Measured by Release of UDP

Some UDP was released from UDP-GlcNAc by each of the zoospore fractions assayed. The distribution of specific activities as assayed here by release of UDP is not inconsistent with that obtained earlier [CAMARGO *et al.*, 1967] as assayed by incorporation of ^{14}C-GlcNAc, insofar as they can be compared (table IX). The specific activity of our gamma particle enriched 20,000 *g* pellet was about 3.5 times greater than that in the whole cell homogenate, while the specific activity of the 5,000 *g* mitochondria enriched pellet was practically negligible. Furthermore, the 20,000 *g* fraction contained nearly 40% of the detectable chitin synthetase in the zoospore.

6. Incorporation of Radioactivity from UDP-GlcNAc During Chitin Synthesis

The time course for incorporation (fig. 36) of radioactivity from UDP-[$^{14}C_1$]-GlcNAc into chitin, measured as chromatographically immobile, labeled material, resembled the response (fig. 35) previously obtained when enzyme activity was assayed via release of UDP, although the former appeared to be more pronouncedly biphasic. The zoospore homogenate incorporated $^{14}C_1$-GlcNAc into chitin at a substantial though decreasing rate for at least 60 min (fig. 36, lower curve). A two-step enrichment of the zoospore homogenate for gamma particles brought about two corresponding increases in the

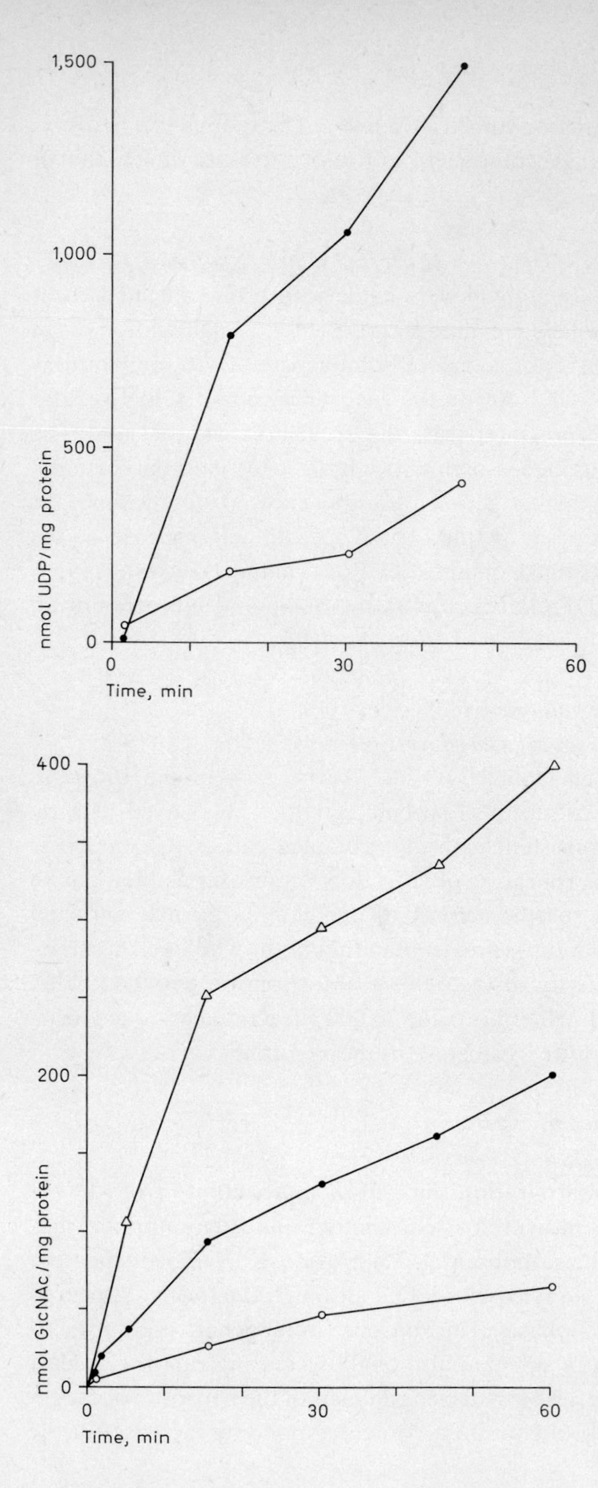

Fig. 35. Release of UDP from UDP-GlcNAc during enzymatic synthesis of chitin by cell-free zoospore extracts. ○ = Whole zoospore sonicate (see chapter VII, section A, 5, a, for preparation); ● = gamma particle enriched 20,000 *g* pellet derived from zoospore homogenate (see chapter VII, sections A, 5, a–c, for preparation). All data points were corrected for endogenous UDP uptake (see chapter X, section B, 9) in the absence of substrate.

Fig. 36. Incorporation of $^{14}C_1$-GlcNAc from UDP-[$^{14}C_1$]-GlcNAc into chromatographically immobile material during enzymatic synthesis of chitin by cell-free zoospore extracts. ○ = Whole zoospore sonicate (chapter VII, section A, 5, a); ● = gamma particle enriched 20,000 *g* pellet derived from zoospore homogenate (chapter VII, sections A, a–c); △ = sucrose density gradient purified gamma particles (chapter VII, sections A, 5, a–d).

activity of chitin synthetase (fig. 36, two upper curves). Both here and in chapter X, section B, 4, the amount of product formed was proportional to the amount of enzyme protein present in our assays.

7. Apparent Association of Chitin Synthetase Activity with Gamma Particles

The high values for specific and total chitin synthetase activities associated with the 20,000 *g* pellet (table IX) suggested that gamma particles contained at least a large proportion of the chitin synthetase in the zoospore. The results (table X) of further purification were consistent with this supposition. Passage of gamma particle fractions through discontinuous sucrose density gradients did not reduce their chitin synthetase activity; rather, as the fractions became more enriched for gamma particles, the specific activity of their chitin synthetase also increased. Up to 19-fold purifications (table IX) were attained for the preparations most highly enriched for gamma particles (see also table X, fn. 2); hence, it seemed as if chitin synthetase was intimately associated with these organelles.

Table IX. Chitin synthesis by different zoospore fractions

Fraction assayed	CAMARGO *et al.* [1967]		Present investigation	
	specific activity[1]	relative activity[2]	specific activity[3]	relative activity
Zoospore homogenate	36	100	280	100
55 *g* (nuclear caps)	1.4	4	–	–
1,000 *g*	2.3	6	55	20
5,000 *g* (mitochondria)	–	–	32	11
10,000 *g*	37	103	–	–
20,000 *g* (gamma particles)	–	–	977	349
20,000 *g* supernatant	–	–	40	14
140,000 *g*	8.5	24	–	–

[1] nmol GlcNAc × (mg protein)$^{-1}$ × (30 min)$^{-1}$.

[2] $\frac{\text{Specific activity of fraction}}{\text{Specific activity of homogenate}} \times 100$.

[3] nmol UDP released from UDP-GlcNAc × (mg protein)$^{-1}$ × (30 min)$^{-1}$.

Table X. Increases in specific activity of chitin synthetase resulting from purification of gamma particle fractions as measured by two different analytical procedures

Fraction assayed	Specific activity[1]		Times purified[2]	
	UDP ↑	$^{14}C_1$–GlcNAc ↓	UDP ↑	$^{14}C_1$–GlcNAc ↓
Whole cell homogenate	280	45	1	1
Gamma particle fraction				
20,000 *g*[3]	977	152	3.5	3.4
Gradient[4]	5340	334	19	7.4

[1] Specific activity is expressed as nmol of UDP released from UDP–GlcNAc or as nmol GlcNAc incorporated from UDP [$^{14}C_1$]–GlcNAc, both × (mg protein)$^{-1}$ × (30 min)$^{-1}$.

[2] These values are based arbitrarily on 30 min assays and, while accurate, they may be too low. By 30 min, the rates of incorporation (Fig. 36) had already decreased; hence, had we used shorter assay times – 15 min for example – the calculated values for «times purified» would have been higher.

[3] Prepared by differential centrifugation of zoospore lysates (chapter VII, sections A, 5, a–c).

[4] Purified by isopycnic sucrose density gradient centrifugation (chapter VII, section A, 5, d).

8. Stability of Chitin Synthetase

The chitin synthetase activities in zoospore homogenates and gamma particle enriched 20,000 *g* pellets were similar in stability when the cell extracts were incubated at 0–4°C in the absence of added substrate before being assayed (fig. 37); rates were essentially stable for at least 120 min. The two preparations did differ in stability, however, when held below 0°C for different time periods. Whole spore homogenates retained full enzyme activity for at least 7 days at –15°C, but the 20,000 *g* pellets lost at least 50% of their activity after only 2 days at –15°C. Further purification of such gamma particle fractions did not enhance the stability of the enzyme under the conditions employed. In fact, when 20,000 *g* gamma particle pellets were further purified on sucrose gradients (chapter VII, section A, 5, d), stored at –15°C, and then assayed by incorporation of radioactivity into chromatographically immobile product, chitin synthetase activity was negligible. Finally, it should be noted that synthetase activity in whole spore homogenates was totally inactivated by a 4 h dialysis at 5°C against 40 mM Tris containing 3 mM EDTA and 10 mM $MgCl_2$ at pH 7.5.

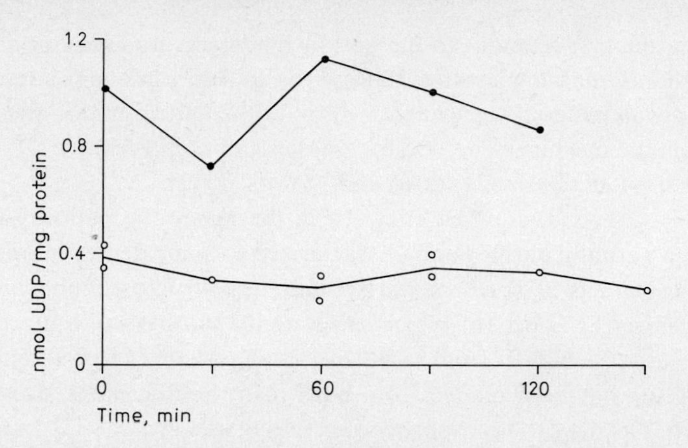

Fig. 37. Stability of chitin synthetase activity in whole zoospore homogenates (○) and gamma particle enriched 20,000 *g* pellets (●) stored at 0 °C in Tris-EDTA buffer at pH 7.5 for different time periods. Chitin synthetase activity was measured via UDP released from UDP-GlcNAc.

9. Reutilization of UDP During Chitin Synthesis

Although UDP formation has not often been monitored as a corroboral tive means of measuring chitin synthetase activity, PORTER and JAWORSK- [1966] did employ this method in their studies of *Allomyces*, as we did in ours with *Blastocladiella*. On the other hand, while they used '...appropriate blanks and UDP standards...', they did not report whether or not UDP was also consumed by their extracts. Our cell-free zoospore extracts did exhibit the capacity to enzymatically metabolize UDP (fig. 38); the time course of its disappearance resembled that for the enzymatic release of UDP (fig. 35) from UDP-GlcNAc. Presumably, this means that UDP might also be reutilized during chitin synthesis; the pathway for its metabolism in *B. emersonii*, however, is not known.

C. Discussion

1. Control of Chitin Synthesis

Growing thalli of *B. emersonii* possess a chitinous cell wall [LOVETT and CANTINO, 1960b], but its zoospores are limited only by a pliable plasma membrane [CANTINO *et al.*, 1963]. However, immediately following flagellar retraction – more appropriately, 'axonemal translocation' – the newly rounded-up spore has already started forming a cyst wall [SOLL *et al.*, 1969; TRUESDELL and CANTINO, 1971]. The wall is synthesized very rapidly, apparently rendering the cyst resistant to disruption by detergents [SOLL *et al.*, 1969]. Although it is not known with certainty[8] if a zoospore contains a reservoir of chitinous wall material somewhere within it, it is known that it does possess the enzymatic machinery needed for synthesis of chitin from UDP-GlcNAc [CAMARGO *et al.*, 1967; CANTINO and MYERS, 1972].

The first, second, and last enzymes in the metabolic pathway to chitin have been partially purified and characterized to some degree in our laboratory using extracts of *B. emersonii*. L-glutamine: D-fructose 6-phosphate aminotransferase (EC 2.6.1.16), which catalyzes the synthesis of D-glucosamine-6-phosphate (GlcN-6-P) from D-fructose-6-phosphate (Fru-6-P) and L-glutamine (Glu), has been refined from both plant homogenates [LOVETT and CANTINO, 1960a, b] and zoospore lysates [NORRMAN *et al.*, 1973]. In the latter, most of the amino transferase activity is in 112,000 *g* supernatants and does not appear to be associated with organelles in the zoospore. On the other hand, chitin synthetase, described by CAMARGO *et al.* [1967] using a 10,000 *g* membrane fraction, has now been shown to be associated with the gamma particles in zoospore lysates. In addition, acetyl-CoA:2-amino-2-deoxy-D-glucose 6-phosphate N-acetyl transferase (EC 2.3.1.4), which mediates formation of N-acetyl D-glucosamine-6-phosphate (GlcNAc-6-P) from GlcN-6-P and acetyl CoA, has been detected in 112,000 *g* supernatants derived from zoospore lysates, separated from the amino-transferase, and purified 400-fold [GIDDINGS and CANTINO, unpublished data]. Also, UDP-N-acetyl-D-glucosamine pyrophosphorylase (EC 2.7.7.f), a uridyltransferase which catalyzes the synthesis of UDP-GlcNAc from UTP and acetyl glucosamine-1-phosphate (GlcNAc-1-P), has been detected in supernatant fractions of zoospore homogenates [unpublished observations cited by CAMARGO *et al.*, 1967]. Thus confronted by a cell which does not actively synthesize chitin yet possesses detectable amounts of almost all the pieces of the enzymatic

[8] CAMARGO *et al.* [1967] have shown that acetylchitodextrins stimulated chitin synthesis in extracts of *B. emersonii*, but interestingly, GlcNAc and diacetylchitobiose were even better actuators. The question still remains, however: Does the zoospore of this fungus initiate formation of entirely new chains of chitin, or does it in fact contain endogenous acceptors of acetylglucosaminyl units? The only published data presently available [LOVETT and CANTINO, 1960b] suggest that zoospores, harvested by procedures designed [MCCURDY and CANTINO, 1960] to prevent encystment, contain about 5 μg of bound glucosamine per mg dry weight which behaved chemically like chitin. However, since the inaugural methods used to keep the spores in a motile state were not tested extensively thereafter [they have since been replaced by more suitable procedures; SUBERKROPP and CANTINO, 1972, 1973], the conclusion that chitin is present in a zoospore must be viewed as tentative and in need of re-examination.

machinery required for it, the question naturally arises: what is the control mechanism that prevents these enzymes from collectively synthesizing chitin until the zoospore encysts?

In attempting to provide explanations for the perplexing enigma that a zoospore contains chitin synthetase but not a chitinous cell wall, CAMARGO *et al.* [1967] considered several possible explanations; one of them involved the chance that some enzymatic step prior to the final transfer of acetylamino-deoxyglucosyl groups might be inoperative. However, since CAMARGO *et al.* [1967] were able to detect UDP-GlcNAc pyrophosphorylase activity in spore extracts, they concluded that the missing enzyme was associated with some earlier step in the metabolic pathway. Now that we have demonstrated the presence of three other enzyme activities that mediate steps which precede the pyrophosphorylase reaction, this sort of explanation is now even less likely than it was then. CAMARGO *et al.* [1967] also suggested that the chitin synthetase in the zoospore may have been present in a physiologically inactive state. Additionally, they discussed a third possible means for its regulation: namely, that since GlcNAc apparently stimulated the chitin synthesizing system [as it does in *Neurospora*, GLASER and BROWN, 1957; *Allomyces*, PORTER and JAWORSKI, 1966; *Mucor*, MCMURROUGH *et al.*, 1971; *Piricularia*, HORI *et al.*, 1971; yeasts, CABIB and BOWERS, 1971] – perhaps by serving as a chain initiator [however, see MCMURROUGH and BARTNICKI-GARCIA, 1971] and possibly, also as an allosteric activator – its presence in the zoospore might function in some regulatory mechanism for *in vivo* chitin synthesis. Allosteric control of chitin synthesis in *Saccharomyces* was also assumed to occur [CABIB and KELLER, 1971] by way of a proteinaceous allosteric effector with unusual stability to heat; however, further work [CABIB and FARKAS, 1971] suggested that this substance was an inhibitor of an activating factor for the chitin synthetase, hence it is no longer tenable to think of it as an inhibitor of the enzyme itself.

Based upon our direct evidence linking chitin synthetase activity to the gamma particle, we would like to propose an alternative means whereby the zoospore might exert a degree of control over the synthesis of chitin. The zoospore is the only stage in the life cycle of *B. emersonii* wherein the gamma particle is found (chapter XII). Immediately prior to encystment, at the time of axonemal translocation, gamma particles begin to decay in a characteristic fashion; this phenomenon includes release of vesicles which migrate to the plasma membrane and an adjacent newly forming cyst wall (chapter III; fig. 6). It is *only* at this time, when microscopically detectable deposition of the cyst wall commences, that gamma particles begin to lose their morphological integrity via this distinctive decay process. These results, and other arguments [CANTINO and MYERS, 1973], are consistent with the premise [TRUESDELL and CANTINO, 1970,1972] that gamma particles bring enzymic and/or structural components to the cell surface and hence are directly involved in the genesis of the cyst wall. We now suggest further that gamma particles contain a large proportion – possibly even all – of the chitin synthetase in a zoospore, and that they serve as barriers which restrict the accessibility of the enzyme to its substrate and thereby suppress its activity. At the time of encystment, the vesicles derived from gamma particles then transport this sequestered chitin synthetase to the cell surface, where the cell wall is synthesized. Our proposal does not rule out the possibility that cytoplasmic inhibitors of chitin synthesis are present in the zoospore, nor does it eliminate possible control by allosteric effectors; it is in fact compatible with both.

An explanation somewhat similar to ours has already been suggested to account for chitin synthetase activity in another fungus, *Mucor rouxii* [MCMURROUGH *et al.*, 1971]. The enzyme was associated with cell walls but also, to a lesser extent, with a 'microsomal' fraction. It was suggested that the chitin synthetase activity in the microsomal fraction was nascent synthetase that had not yet migrated to the cell wall. However, neither the identity nor origin of the chitin synthetase containing structures in the microsomal pellets was established.

Our proposal that much if not all of the chitin synthetase in the zoospore of *B. emersonii* is contained in its gamma particles is a reasonable extrapolation of the fact (fig. 36) that it is progressively enriched as gamma particles are increasingly purified; it is also in keeping with the early postulation [DE DUVE and BERTHET, 1954] – since found to be fairly applicable to many enzyme activities – that certain enzymes are associated with specific subcellular components.

We will not expand here upon the much broader but cardinal question: by what means is encystment as a whole – and therefore also decay of the gamma particle and chitin synthesis – prevented from occurring in a non-encysting zoospore to begin with? The interested reader may wish to consult a brief essay on this matter of causality in a recent review [TRUESDELL and CANTINO, 1970, pp. 39–43].

2. Gamma Particles do not Appear to be Lysosomes or Microbodies

Numerous enzyme activities have been associated with preparations of cell organelles limited by unit membranes [DE DUVE, 1969], and this has led to the evolution of a biochemical taxonomy for this class of particles (peroxisomes, glyoxysomes, lysosomes, microbodies, etc.). In particular, lysosomes

[DE DUVE *et al.*, 1955] are relatively electron-dense organelles limited by a single unit membrane [see FAWCETT, 1966, for illustrative examples] which contain hydrolytic enzymes [DE DUVE and WATTIAUX, 1966; DE DUVE, 1969]. One enzyme apparently present in all lysosomes studied thus far is acid phosphatase. Not only do gamma particles differ from lysosomes in their internal structure (chapter III), but they also fail to display either acid or alkaline phosphatase activities.

Microbodies, which occur in the cells of a limited number of higher animal tissues [HRUBAN and RECHCIGL, 1969] and are limited by a single unit membrane, often contain an electron-dense core. Microbodies are also found in numerous higher plant cells [MOLLENHAUER *et al.*, 1966; FREDERICK *et al.*, 1968; TOLBERT and YAMAZAKI, 1969; VIGIL, 1970; HUANG and BEEVERS, 1971] and some lower organisms [BRACKER, 1967; MÜLLER, 1969]; some of them also include electron-dense nucleoids. Catalase is present in almost all microbodies [HRUBAN and RECHCIGL, 1969] – although apparently not without exception [e.g. the glyoxysomes in *Euglena*, GRAVES *et al.*, 1972] – including glyoxysomes [HUANG and BEEVERS, 1971] and peroxisomes [FREDERICK and NEWCOMB, 1969]. Here too, gamma particles differ from microbodies not only on the basis of their internal morphology, but also because they lack demonstrable catalase activity.

In conclusion, gamma particles lack acid and alkaline phosphatase as well as catalase activities but, on the contrary, do possess chitin synthetase activity; they contain DNA and RNA (chapters VIII and IX); they have an exceptionally high buoyant density (chapter VII); they exhibit a markedly ordered and distinctive morphology; and they decay by a characteristic mechanism (chapter III). We are left with the inescapable conclusion that these organelles should not be classified as either lysosomes or microbodies.

3. Relationship Between the Activities of the Gamma Particle in B. emersonii *and Mechanisms of Cell Wall Formation in Other Fungi*

To conclude that participation of cytoplasmic vesicles in fungal cell wall formation has now '...actually been established...' [MIMS, 1972] is perhaps an overstatement. However, strong *correlations* do exist in support of the hypothesis [MCCLURE *et al.*, 1968; BRENNER and CARROLL, 1968; GIRBARDT, 1969; GROVE *et al.*, 1970; TRUESDELL and CANTINO, 1970; GROVE and BRACKER, 1970; BRACKER, 1971; HEATH *et al.*, 1971; MARCHANT and MOORE, 1973]. Fungi contain vesicles which migrate to various sites on a plasma membrane, apparently fuse with it, and perhaps liberate their supposed contents (membrane material? wall material? wall precursors? enzymes? etc.) into a region nearby; a pre-existing cell wall may – but need not – already be present next to the plasma membrane when these events occur. But, there is not yet available any hard evidence – direct chemical or enzymatic evidence – to support the hypothesis. Clearly, there is a need to isolate undamaged fungal vesicles and to characterize them biochemically if the validity of the supposed role that these vesicles play in cell wall formation is to be established. To date, no such results have been published.

The immediate progenitors of such vesicles in *B. emersonii*, i.e. the gamma particles, have been isolated and shown to contain chitin synthetase. As already emphasized, gamma particles decay in a distinctive fashion during zoospore encystment with concomitant release of vesicles which migrate to the cell surface. The available facts support the premise that the chitin synthetase sequestered in the gamma particle is involved in the *de novo* genesis of a cyst wall: (a) the spore surface is the site – and the *only* site – of demonstrable [SOLL *et al.*, 1969; TRUESDELL and CANTINO, 1971] wall deposition; (b) the initial wall is not detectable microscopically until after the vesicles derived from gamma particles begin to accumulate at such sites [TRUESDELL and CANTINO, 1970, 1971]; and (c) gamma particles only – and always – exhibit their characteristic decay process, with its release of vesicles, in spores that are laying down a cyst wall [TRUESDELL and CANTINO, 1970, 1971].[9] *B. emersonii* zoospores do not possess either an endoplasmic reticulum (as commonly defined) or typical dictyosomes. However, the *origin* of gamma particles may involve a relationship with a ribosome-laden membrane system (chapter V). Hence, in view of the origin of secretory vesicles from dictyosomes in some Oomycetes [GROVE *et al.*, 1970], the gamma particle may be the functional equivalent of a dictyosome in the sense that it generates vesicles which mediate the export of materials – presumably, in this instance, of cell wall building enzymes – to the surface of the cell during encystment.

[9] This last argument finds corroboration in an apparent foreglimpse, 20 years ago, of things to come; at that time we had tabulated evidence [CANTINO and HYATT, 1953a], table II, that most of the *orange* zoospores of *B. emersonii* (i.e. those derived from O plants as depicted in figure 1) are nonviable, and we wrote (p. 35) about them in these words: 'Usually no germination occurs. The flagellum is often retained and remains visible on the agar, a phenomenon which seldom occurs with swarmers from OC or RS plants...It is proposed that inability to retract the flagellum...is in keeping with the nonviable character of most swarmers from O plants.' In the light of the more recent observations that O spores possess relatively few gamma particles (fig. 2), this seemingly prophetic suggestion is now seen to be at least consistent with, if not actually additional evidence in support of, our conclusion [TRUESDELL and CANTINO, 1971] that gamma particles are intrinsically required in the normal encystment process.

XI. Effect of Visible Light on Gamma Particles

Visible light, which induces a diversity of effects in growing plants of *B. emersonii* [CANTINO, 1965], also causes the zoospores derived from them to encyst [CANTINO and TRUESDELL, 1971b]. In particular, when a suitably dense zoospore suspension ($\geqq 5 \times 10^5$ spores/ml) derived from OC plants grown in the dark is incubated in the dark, few zoospores exhibit the capacity to encyst; however, non-encysting members of this suspension *are* triggered to encyst if they are exposed to visible light. Such a light induction is insensitive to cycloheximide. This type of encystment has a mean time at 22°C of 320 min, and it ceases abruptly if the zoospore suspension is transferred back into the dark at any time before all the zoospores with the capacity to encyst in response to light have done so. This phenomenon has been labeled [CANTINO and MYERS, 1972] the 'direct effect' of light. However, when OC plants are grown under light of appropriate intensity, the zoospore populations derived from them do not display the foregoing light-induced encystment, i.e., they encyst equally well in light and dark; this has been labeled the 'indirect effect' of light [CANTINO and MYERS, 1972]. Since gamma particles, by virtue of both their function and their enzyme content, had seemingly become intrinsically entwined in the process of encystment, attempts were made to determine if they, too, were affected by light.

A. Methods and Materials

1. Culture Methods and Preparation of Zoospore Suspensions

OC plants of *B. emersonii* were grown on PYG agar and incubated at 22–24 °C either in the dark or under 240 μW/cm² of 'cool white' fluorescent light. Zoospore populations were produced in the usual way by flooding the culture plates with H_2O.

2. Electron Microscopy

Zoospores were fixed and stained in a solution of 0.25 % OsO_4 and 0.09 % Ur acetate in 4.8 mM Na cacodylate [TRUESDELL and CANTINO, 1970; fixation II].

3. Staining and Counting of Gamma Particles in situ

Gamma particles were stained by slightly modifying the procedure of MATSUMAE *et al.* [1970]. Prechilling the zoospores at 0–1 °C prior to fixation caused them to expand [SHAW and CANTINO, 1969], thereby improving the contrast, and hence, the resolution of stained gamma particles when viewed by light microscopy. Accordingly, spores were chilled for 30 min, fixed in 0.4% glutaraldehyde for 15 min at 0–1 °C, air dried on glass slides, and exposed to aqueous methylene blue at various concentrations $\leq$ 0.25%, depending on the desired speed of the staining reaction; alcoholic solutions of the dye were not used because ethanol caused cold-expanded, glutaraldehyde-fixed spores to shrink again.

4. Isolation of Gamma Particles and Assay for Chitin Synthetase Activity

Gamma particle suspensions were prepared from zoospore homogenates by isopycnic sucrose density gradient centrifugation of 20,000 *g* pellets (chapter VII, section A, 5, d) and sonicated for 30 sec with 75 W at 0–1 °C before being assayed. Chitin synthetase was estimated by following incorporation of $^{14}C_1$-GlcNAc into chromatographically immobile reaction product (chapter X, section A, 4). Protein was determined according to LOWRY *et al.* [1951].

5. Methods for Evaluating Zoospore Encystment

Zoospore suspensions were gently passed through filter paper about 15 min after culture plates were flooded with H_2O and then, without further manipulation, immediately put into 50 ml Erlenmeyer flasks at 22 °C and shaken at 76 oscillations/min, either in the dark or over 3,200 μW/cm² of incandescent light. Samples were removed at intervals, fixed and stained as described above, and scored for numbers of gamma particles per spore. Simultaneously, encystment was followed using counting chambres [CANTINO and TRUESDELL, 1971b] and analyzed by transforming typical sigmoid encystment curves into linear time-encystment functions [TRUESDELL and CANTINO, 1971].

B. Results

1. Light-Induced Reduction in the Number of Gamma Particles per Zoospore: An Event Associated with the 'Indirect Effect'

Zoospores examined immediately after harvest from light-grown OC plants had less than half the number of gamma particles (mean: 5.4) ordinarily found in zoospores derived from dark-grown plants (table XI). The frequencies for gamma particles per spore in both instances were normally distributed about their means (fig. 39).

2. Light-Induced Change in Zoospore Structure: Genesis of Cytoplasmic Granules Associated with the 'Indirect Effect'

Light did not visibly alter any of the spore's internal structures (see chapter III for zoospore organization). But, light did cause a new structure to

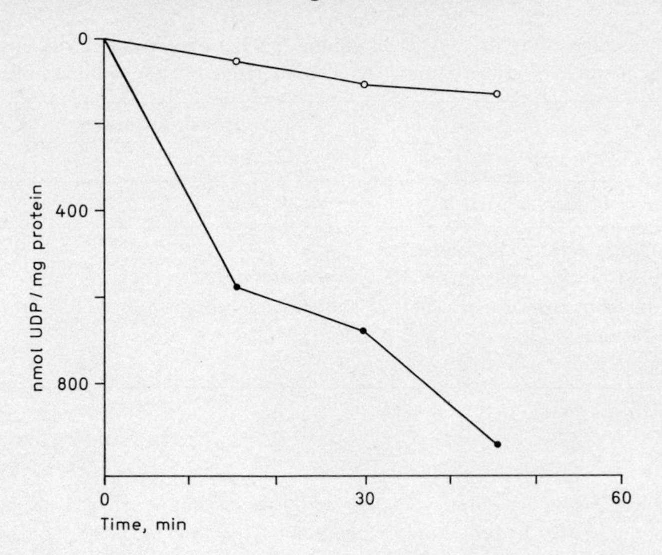

Fig. 38. Disappearance of added UDP in the presence of cell-free zoospore extracts. ○ = Zoospore sonicate (see chapter VII, section A, 5, a, for preparation); ● = gamma particle enriched 20,000 *g* pellet derived from zoospore homogenate (see chapter VII, sections A, 5, a–c, for preparation).

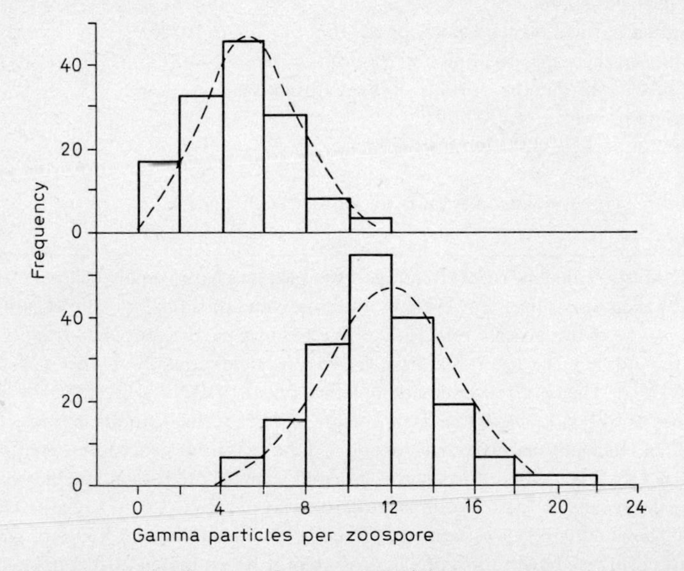

Fig. 39. Frequency distribution of gamma particles per zoospore. Upper and lower histograms correspond to means in table XI for zoospore derived from light- and dark-grown plants, respectively. The normal distribution curves were fitted to the histograms and tested for fit by standard statistical procedures.

Table XI. Difference in mean number of gamma particles in zoospore suspensions[1] derived from light and dark-grown plants: a result associated with the «indirect effect» of light

	Growth conditions	
	light	dark
Mean number of gamma particles	5.4[2]	11.8[3,4]

[1] Population density: 10^6 spores/ml.

[2] Results from five experiments; 25 spores were scored in each.

[3] Results from eight experiments; 25 spores were scored in each. For other statistics and related details, see CANTINO and MYERS [1972].

[4] Compare with mean values in figure 4.

Table XII. Representative chitin synthetase activities in gamma particle suspensions prepared by procedure E (chapter VII, sections A, 5, a–c)

	Spores derived from	
	dark-grown OC plants	light-grown OC plants
Gamma particles/spore	11.8	5.4
Total gamma particle protein/spore, pg $\times 10^3$	220	270
Total chitin synthetase/spore, nmol $\times$ 30 min^{-1} $\times 10^8$	117	111
Specific activity of gamma particle chitin synthetase, nmol $\times$ (mg protein)$^{-1}$ $\times$ (30 min)$^{-1}$	152	90
Total gamma particle chitin synthetase/spore, nmol $\times$ 30 min^{-1} $\times 10^8$	27.9	11
Total chitin synthetase/gamma particle, nmol $\times$ 30 min^{-1} $\times 10^8$	2.4	2.0

[1] Attempts to calculate recoveries of gamma particles based on direct counts are difficult because of their small size. However, by making some presumably valid assumptions, an estimate can be made, at least for the 20,000 *g* gamma particle enriched fractions. On the basis of the indole DNA test (table IV.), gamma particles contain about 1.8% of the total zoospore DNA. The contribution made by component IV DNA (fig. 25) to the total DNA absorbance is about 1.5%. If the assumption is correct that gamma particle DNA was extracted as efficiently as the other zoospore DNAs by the procedures used, and if the amount of DNA per gamma particle is normally distributed, then the close agreement betweeen the results of these two independent assays (1.8 vs. 1.5%) suggests that most if not all of the gamma particles were recovered. Possible subsequent losses of gamma particles during further purification procedures cannot be estimated at this time.

Although the foregoing argument rests upon analyses of spores from dark-grown plants, preliminary studies suggest that comparable recoveries of gamma particles are obtained from spores derived from light-grown plants.

appear: numerous aggregates of electron-opaque granules in various sizes were dispersed in the cytoplasm of the spores from light-grown plants (fig. 40b, d hollow arrows). These granules apparently had the same characteristic density as the matrix in a gamma particle (fig. 40b, d, solid arrows). However, in corresponding sections (fig. 40 a, c) through zoospores derived from dark-grown plants, there were few if any such granules.

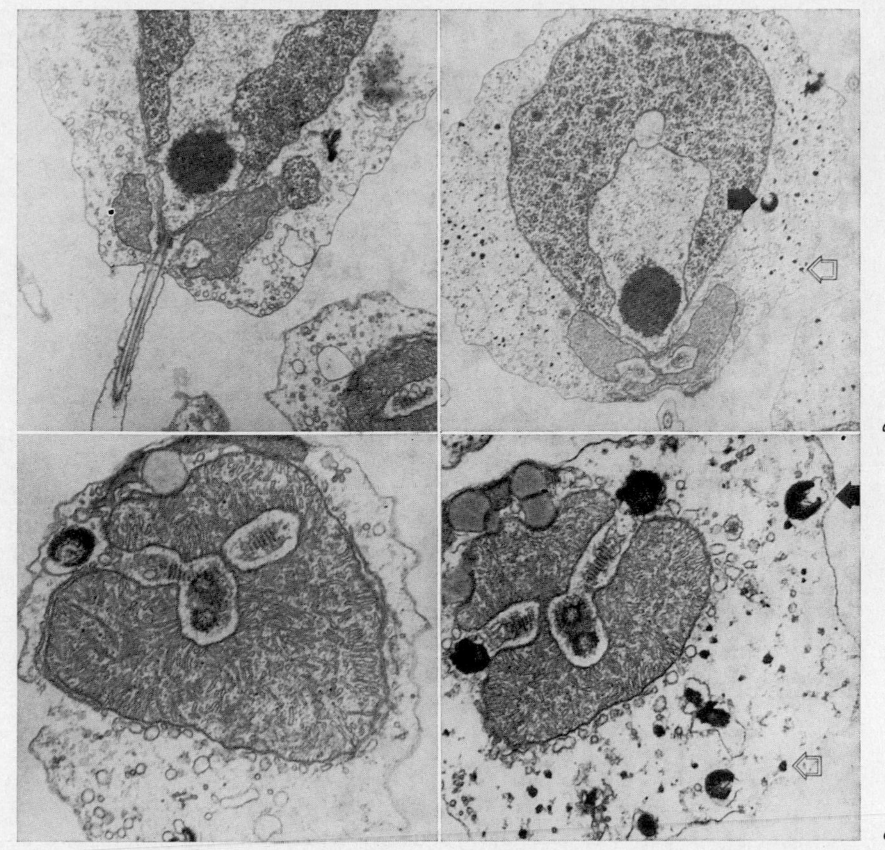

Fig. 40. Median longitudinal sections (a, b) and transverse sections through the rootlet-kinetosome-centriole region (c, d) of zoospores derived from OC plants grown in the dark (a, c) and under 240 μW/cm² of light (b, d). Note dark granules (hollow arrows) and gamma particles (solid arrows). *a* × 6400, *b* × 6300, *c* × 11300, *d* × 14000. From CANTINO and MYERS [1972].

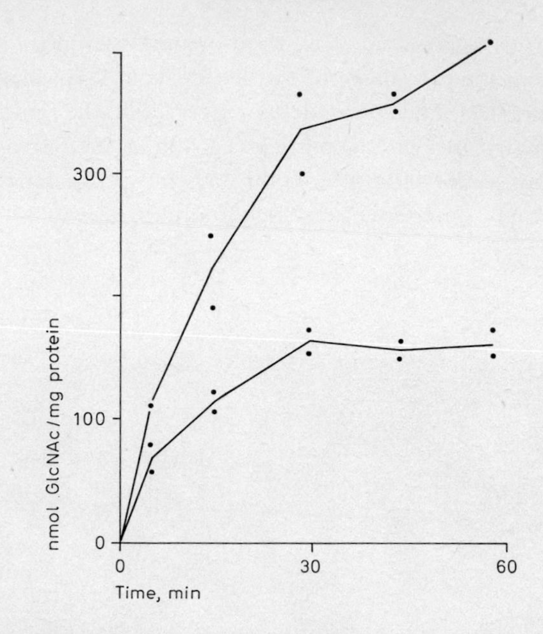

Fig. 41. Chitin synthetase activity in gamma particles from zoospores derived from light- (lower curve) and dark-grown (upper curve) plants.

Table XIII. Mean number of gamma particles in zoospores derived from dark-grown plants and incubated either under 3,200 μW/cm² of incandescent light or in the dark

	Mean number of gamma particles/spore after incubation	
Time, h	light[1]	dark
0	11.8[2]	11.8[3]
4	6.8	6.9
8	5.5	5.7

[1] Spore populations were of suitable population density to exhibit light-induced encystment [CANTINO and TRUESDELL, 1971].

[2] Each mean value in the table is the result of eight experiments, 25 spores being scored in each. Tests of significance, other statistics, and additional details about these experiments are provided by CANTINO and MYERS [1972].

[3] The same zero time spores were used for both series.

3. Light-Induced Alteration in Activity of Gamma Particle Chitin Synthetase: A Change in Enzyme Distribution Associated with the 'Indirect Effect'

The chitin synthetase in gamma particles from spores derived from dark-grown plants mediated incorporation of $^{14}C_1$-GlcNAc into chromatographically immobile material at a substantial, albeit decelerating, pace for at least 1 h (fig. 41, upper curve). However, the chitin synthetase in gamma particles from spores of plants grown under about 250 $\mu W/cm^2$ of light fixed labeled GlcNAc less rapidly and only for about 30 min (fig. 41, lower curve). The specific activity of the synthetase from light-grown plants was only 59% of that from dark-grown plants (table XII). Similarly, the total gamma particle chitin synthetase per spore from light-grown plants was only 40% of that from dark-grown plants (table XII). These results correlate well with the 46% decrease in the number of gamma particles per spore (table XI) induced by the 'indirect effect' of light.

4. Apparent Absence of Light-Induced Changes in Gamma Particles During the 'Direct Effect' of Light on Zoospore Encystment

The mean time of light-induced encystment is 320 min; this is much greater than the mean time of about 15–30 min for encystment induced by other means [CANTINO and TRUESDELL, 1971b]. Consequently, only about 10–20% of the zoospores (derived from dark-grown plants) *destined to encyst* because of the 'direct effect' of light had actually done so after only 4 h of incubation in the light (e.g. in CANTINO and TRUESDELL, 1971b, fig. 1 or CANTINO and MYERS, 1972, fig. 2]. However, light did not simultaneously alter the number of gamma particles per spore in the same population. For example, in the non-encysted zoospores that remained after 4 h of incubation in the light, the number of gamma particles per spore had been reduced to about half the starting level (table XIII). However, among zoospores incubated in the dark for 4 h, the number of gamma particles per spore was almost identically reduced (table XIII). Furthermore, although the number of gamma particles per spore decreased slightly after an additional 4 h of incubation, this second decrease was again about the same for light- and dark-incubated spore suspensions.

C. Discussion

Zoospores from light-grown OC plants, when scored immediately after being harvested, contain less than half the number of gamma particles found in zoospores derived from dark-grown plants. The former zoospores, but not the latter, contain numerous aggregates of strongly osmiophylic cytoplasmic granules with an electron opaqueness that resembles that of the matrix in a typical non-decaying gamma particle. This could mean that the osmiophylic granular material is a normal component of gamma particles, and that visible light either (a) prevents its incorporation into gamma particles before or during sporogenesis, or (b) triggers its release from gamma particles immediately after sporogenesis, or, at least, in the short interval before the spores are discharged. Whichever is the case, the 'indirect effect' of light on zoospore encystment also results in fewer identifiable gamma particles in the average zoospore.

The data illustrating the effect of visible light on chitin synthetase activity provide direct evidence linking a light-induced increase in encystment capacity to an enzyme intimately associated with the encystment process. Gamma particles derived from zoospores obtained from dark-grown plants incorporate $^{14}C_1$-GlcNac into chitin more rapidly and for longer periods than those obtained from light-grown plants. Hence, light seems to have a pronounced effect on the activity of gamma particle chitin synthetase. The data in table XII offer further insight into this phenomenon. Although the number of gamma particles per spore from light-grown plants is only about half of that from dark-grown plants, the protein associated with the total number of gamma particles per spore is almost the same for both. Therefore, it follows that the average gamma particle in zoospores from light-grown plants must have contained twice as much protein as the average gamma particle in zoospores from dark-grown plants. The specific activity of gamma particle chitin synthetase in zoospores from light-grown plants is about half of that from dark-grown plants, a difference also reflected in the total gamma particle chitin synthetase activities per zoospore. However, if enzyme activities are compared on the basis of the total chitin synthetase per gamma particle, this difference is no longer evident; irrespective of lighting conditions, all gamma particles carry the same amount of enzyme. In conclusion, visible light apparently affects the incorporation of chitin synthetase into gamma particles – either by limiting the amount that is confined in them when zoospores are made, or by causing some of the enzyme to be released from gamma particles before spores are discharged. Yet, in either case it does not significantly alter the total amount of the synthetase in the zoospore.

As for the 'direct effect' of light, the decrease in the number of gamma particles while zoospores are incubating in either light or dark deserves brief comment. First, it might logically be argued that in any suspension of zoospores, those with the greatest number of gamma particles would be the first ones to encyst, thus enriching the population with those zoospores containing fewer gamma particles. Accordingly, the measurable gamma particle content of the remaining non-encysted spores (i.e. as in table XIII) would not really reflect the gamma particle content of the population as a whole. However, this possibility is unlikely for reasons spelled out elsewhere [Cantino and Myers, 1972]. Second, there is the important question: What happens to the gamma particles that vanish in zoospores swimming about over an 8 h period in either light or darkness? At this time, judging mainly from electron-microscopic observations of zoospores kept swimming for up to 20 h [Suberkropp and Cantino, 1973], we can only conclude that the gamma particles in such zoospores do not seem to 'decay' in the sense that this term denotes [Truesdell and Cantino, 1970, 1971] the relatively fast process by which gamma particles change characteristically in structure and generate membranous elements during encystment.

Finally, there is one other aspect of the 'direct effect' of light not touched upon thus far which should be mentioned in closing. Previously published data [Cantino and Truesdell, 1971b] show that, if freshly harvested zoospores derived from dark-grown plants are kept in the dark for varying periods of time before being illuminated, the mean time of encystment for such zoospore populations is inversely related to the length of time the spores have been preincubated in the dark. We have interpreted this to mean that such zoospores must be altered in some way before they can exhibit light-induced encystment, and have proposed [Cantino and Myers, 1972] that the demonstrable evanescence of gamma particles during these time intervals constitutes part, if not all, of this essential alteration. Since it occurs equally in light- and dark-incubated spore suspensions (table XIII), it can logically be called a 'dark reaction'. And because light is needed continuously [Cantino and Truesdell, 1971b] for light-induced encystment, the 'dark reaction' probably precedes the light reaction and presumably is required for it. We have tried to bypass the radiant energy requirement for light-induced encystment by providing zoospores being held in the dark with low concentrations of metabolic intermediates on the path to chitin synthesis; asterisks denote the compounds tested thus far:

```
*Glc–6–P
   ↓                                                        *UTP
*Fruc–6–P                                                     ↓
            ) → *GlcN–6–P → GlcNAc–6–P → GlcNAc–1–P → *UDP–GlcNAc.
*Glutamine                      ↑                               ↓
   ↑                          *GlcNAc                         Chitin
*Glutamate
```

Table XIV. The capacity of metabolic intermediates in chitin synthesis to function as substitutes for radiant energy in the «direct effect» of light

Metabolite	Concentration, mM	Effectiveness[1], %
Glucose-6-phosphate	0.45	+3
Fructose-6-phosphate	0.45	0
Glutamine	0.45	–8
Glutamate	0.45	+11
Glucosamine-6-phosphate	0.45	–17
Glucosamine-6-phosphate	0.03	– 2
N-acetyl glucosamine	0.45	0
Uridine triphosphate	0.45	+20
UDP-N-acetyl glucosamine	0.45	+146
UDP-N-acetyl glucosamine	0.08	+ 36

[1] In each experiment, final percent encystment (E_{max}) in the light control was corrected for E_{max} in the dark control; the capacity of a compound to substitute for light was then estimated as follows:

$$\frac{(E_{max} \text{ with metabolite}) - (E_{max} \text{ dark control})}{(E_{max} \text{ light control}) - (E_{max} \text{ dark control})} \times (100) = \% \text{ effectiveness.}$$

Therefore, a +% represents stimulation and a –% represents inhibition relative to the response to light, which is set at 100% effectiveness.

The results obtained (table XIV) – although preliminary in that most experiments have only been done in duplicate – suggest that UDP-GlcNAc is the only metabolite on this pathway that can substitute for light. Consequently, as a partial basis for our on-going studies, we are hypothesizing that the 'direct effect' of light on encystment of *B. emersonii* zoospores involves the following sequence: (1) gamma particle $\xrightarrow{\text{dark}}$ X; (2) $\xrightarrow{\text{light}}$ UDP-GlcNAc; (3) $\xrightarrow{\text{dark}}$ encystment. The first step is a 'dark reaction' involving a breakdown of some of the gamma particles with concomitant release of unidentified material, X. The second step is a light-dependent reaction involving conversion of unknown substances (possibly related to X) to UDP-GlcNAc. The final step, also a 'dark reaction', is encystment itself.

In almost all of the work being reported in the recent literature on fungal zoospores, either the lighting conditions to which the spores are exposed have not been indicated or, if arbitrarily specified, the relative effects of light and darkness on the observations being made have not been tested. It is surprising that more attention is not being given to this potential variable; it should not be unexpected, therefore, if some of the interpretations of results published thus far about the zoospores of various fungi have to be re-evaluated when the effects of visible light have been taken into consideration.

XII. Concluding Remarks

In this monograph, we have tried to bring together what is currently known about the character of the gamma particle with a stereoscopic interpretation of the configuration[10] of its skin and backbone, a partial inventory of its macromolecular cargo, and an account of its *modus operandi* during zoospore differentiation. We have seen that gamma particles arise *de novo* by way of a coalescence of tiny electron-dense granules bound by flattened cisternae in the cytoplasm of a differentiating sporangium, and that later they are released from the plant within its zoospores. Once methods had been perfected for isolating and purifying gamma particles, an interesting threesome of chemical ingredients was detected associated with them: DNA-IV, the lightest of the spore's three DNA satellites, an RNA of low molecular weight, and a chitin synthetase. But then, almost as if this not inharmonious assemblage was destined to be rendered a more tangled skein, it also turned out that the intracellular distribution of the chitin synthetase – not to mention the behavior of the gamma particle and the zoospore itself – was photosensitive.

Eventually, just as the life history of a zoospore must perforce come to a close once it commences to encyst, so too the life history of the gamma particle is obliged to end as it begins to decay; in the process, it releases vesicles which, we presume, carry its chitin synthetase to the surface of the spore. Accordingly, we re-emphasize here the salient point about which we wish to make some final comments. Initially, the matrix of a gamma particle grows out of the amalgamation of small granular 'subunits'. Later, the matrix loses its identity as it decays into small vesicular 'subunits', and derivatives of these subunits become intrinsically involved in synthesis of the cell wall material that accompanies encystment.

[10] We have searched for plausible reasons – even somewhat fanciful ones at times – why the curious shape of a gamma particle might possibly provide either phylogenetic or ontogenetic survival value for the *Blastocladiella* zoospore; thus far, our efforts to find the *raison d'être* have been in vain.

A. Function of the Gamma Particle in a Swimming Zoospore

The zoospore of *B. emersonii* does not synthesize RNA [LOVETT, 1968; SCHMOYER and LOVETT, 1969; SOLL and SONNEBORN, 1971b] or protein above barely detectable amounts [SOLL and SONNEBORN, 1971a]. It does, however, become biosynthetically active upon encystment [LOVETT, 1968; SOLL and SONNEBORN, 1971a, b]. To suppress these and related processes, this water fungus has evolved – or so it would seem from our preconditioned viewpoints – protective devices for safeguarding supposedly important components within its zoospores until they are required for growth and development. One such device is the nuclear cap, which may function to protect intact ribosomes until they are required during germination [LOVETT, 1968; SCHMOYER and LOVETT, 1969]. Another is the gamma particle, which may serve not only as a protective package for sequestering other kinds of key ingredients but also, perhaps, as an instrument for transferring them from one generation to the next by way of the organism's only reproductive vehicle, the zoospore. These particular ingredients may not only have special roles to play, they may also have to go into action more or less simultaneously as soon as encystment begins; if so, it would stand to reason that the gamma particle might provide exclusive protection for them until needed.

B. Related Questions of Gamma Particle Autonomy and Continuity, Significance of its DNA, and Replication of its 'Subunits'

To be genetically autonomous by contemporary standards [TEWARI, 1971], a gamma particle would have to contain a stable carrier of genetic information capable of being replicated, DNA-dependent RNA polymerase, and protein synthesizing machinery (i.e., ribosomes, mRNAs, tRNAs, and amino acid activating enzymes). In addition, there should be evidence that it contains a DNA polymerase and that the DNA can specify some structural and functional attributes of the gamma particle.

Gamma particles are not formed *via* pre-existing images as templates. In this sense, they are not 'self-reproducing'; they do not exhibit structural continuity. Rather (fig. 42), they arise *de novo* during sporogenesis [LESSIE and LOVETT, 1968; BARSTOW and LOVETT, 1974] by coalescence of dense granules, sheets of cisternae, and perhaps other things for the once-in-a-life-cycle reproduction of the gamma particle. The process is in phase (fig. 43) with the final round of nuclear divisions at the end of the organism's generation time.

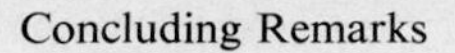

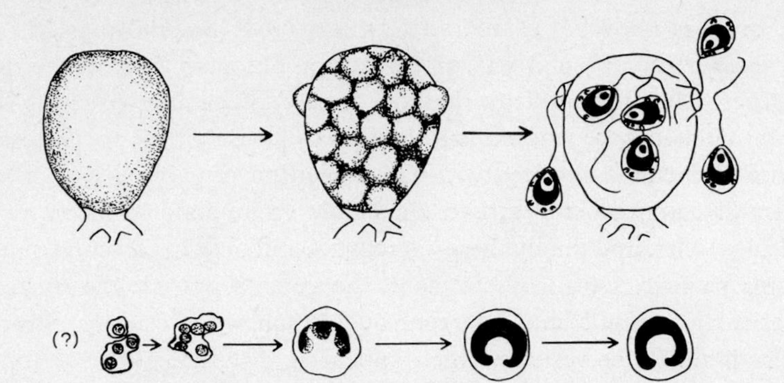

Fig. 42. Gamma genesis during sporogenesis in *B. emersonii.*

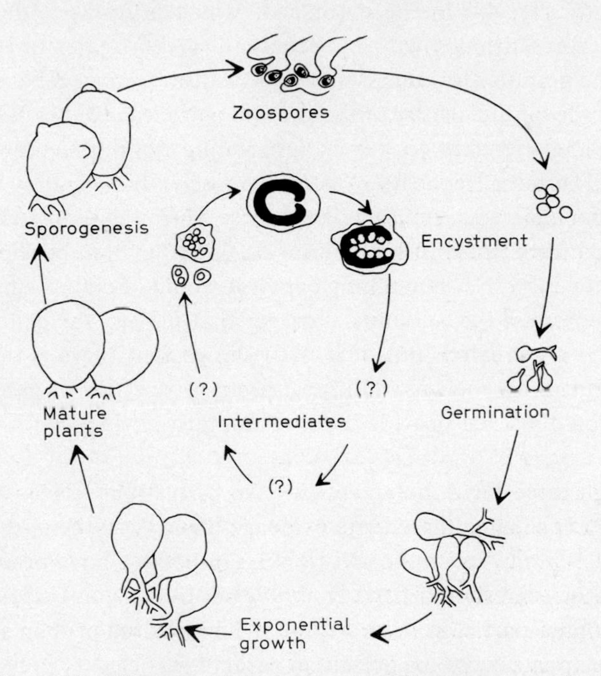

Fig. 43. The life cycles of *B. emersonii* and its gamma particles.

Thus, the number of gamma particles per spore may be a function of the available granules at the time of sporogenesis. It can reasonably be supposed that it will one day be possible to manipulate environmental conditions in such a way as to upset this apparent connection between 'subunits' and gamma particles and, thereby, establish the reality of its cause-and-effect relationship.

As far as is known [LESSIE and LOVETT, 1968; BARSTOW and LOVETT, 1974], these granular 'subunits' appear during ontogeny only at the time of sporogenesis. It is noteworthy that the 'decay' of gamma particles occurs only during zoospore encystment and that it resembles in reverse, the sequence of events associated with the formation of gamma particles. The granules first appear in cisternae, then they accumulate in larger roughly spherical vesicles, and finally these aggregates condense to form the matrix of a gamma particle. During encystment, the gamma particle matrix releases tiny vesicular 'subunits' into a larger, roughly spherical vesicle, and then the latter buds off other vesicles which appear in cisternae in the cytoplasm. It does not seem to be unreasonable, therefore, to postulate that there may exist another, smaller, heretofore unrecognized kind of cytoplasmic 'intermediate' (fig. 43) in the cytoplasm which is released during encystment, replicates during growth, and eventually gives rise to – or is incorporated into – the granular 'subunits' that appear during sporogenesis. The result would be a sort of life history of the gamma particle, and it is illustrated here in temporal relationship to the corresponding developmental cycle of *B. emersonii*. The term 'replication' is used loosely. Justification for it would require proof that new 'intermediates' arise during exponential growth by being copied from pre-existing 'intermediates'. If it could be shown that this 'intermediate' is a DNA-containing entity, it would of course still not explain by what *mechanism* the 'subunits' – or, for that matter, the gamma particles themselves – are created. But, if it were shown that there is present in the growing plant a cytoplasmic DNA with the properties of gamma particle DNA, it would at least provide some presumptive evidence for genetic continuity.

Although three RNA polymerases have been found [HORGEN, 1971] in zoospore homogenates, there is no evidence that any of them resides in gamma particles. Similarly, aminoacyl-tRNA synthetases have been detected in zoospores [SCHMOYER and LOVETT, 1969] but there is no evidence that they occur in gamma particles. In short, the machinery for protein synthesis has not been demonstrated to be present in gamma particles. However, the fact is that the search for it has simply not been made. On the other hand, gamma particles do contain tRNA. If this tRNA differs from other tRNAs in *B. emersonii* zoospores, competitive hybridization experiments might yield evidence that gamma particle DNA has a coding function. Furthermore, if our speculation that gamma particle DNA may contain codons for chitin synthetase turns out to be correct, then there should exist a species of mRNA specific for this enzyme which will also hybridize with gamma particle DNA.

In conclusion, the gamma particle appears to be interpositioned on the path between an apparently autonomous state in the zoospore and a state as yet unexplored in the growing plant. It now becomes especially important to establish two things in particular: (a) whether or not there is genetic continuity between the granular 'subunits' that serve as building blocks in the synthesis of a gamma particle during sporogenesis, and the vesicular 'subunits' into which it decays during zoospore encystment, and (b) to what extent, if any, gamma particles determine their own replication, i.e. to what degree the nucleus and the gamma particles in a zoospore are independent or interdependent informational units.

The gamma particle provides a means of investigating the regulatory role of a relatively new, well-defined, subcellular organelle in the interlocking control mechanisms of a cell – a dual role, whereby the apparently limitative function of a gamma particle in a dormant non-differentiating cell (i.e. a zoospore) is replaced by an export function once the cell is triggered to encyst, that is, to differentiate.

When compared to what has been learned about mitochondria and chloroplasts – an accomplishment resulting from the combined efforts of many laboratories over many years – our present knowledge of the gamma particle is still in its infancy. We would like to think, however, that we have helped to crack open the door to more extended investigations of this interesting and structurally distinctive organelle. In particular, it promises to become, in our opinion, a useful experimental system for studying the function of a small piece of unanchored genetic material in the cytoplasm of a well-characterized, time-tested fungus. We have every expectation that other laboratories will now become involved in this effort and that, thereby, in the years to come the gamma particle may take a place, along with mitochondria and plastids, as a third important vehicle for investigations of extrachromosomal heredity.

C. Does the Gamma Particle Belong to a Family of Widespread Cytoplasmic Organelles Involved in Encystment of Fungal Zoospores?

In a preliminary attempt to assess whether or not the gamma particle in *B. emersonii* might be but a morphologically specialized form of a subcellular structure ubiquitously distributed among zoospores of the various species of water fungi, TRUESDELL and CANTINO [1970] wrote: 'Encystment is a universal process among the zoospores of aquatic Phycomycetes. If gamma particles play an integral role in this phenomenon, might they also belong to some class of organelles that occurs universally in fungal zoospores? When the appearance of a gamma particle is defined in general terms, i.e., as an electron dense particle contained in a vesicle, then similar structures do indeed occur in the motile cells of species of *Allomyces* [BLONDEL and TURIAN, 1960; FULLER and CALHOUN, 1968; MOORE, 1968] *Monoblepharella* [FULLER and REICHLE, 1968], *Rhizidiomyces* [FULLER and REICHLE, 1965], *Rhizophlycitis* [CHAMBERS and WILLOUGHBY, 1964], *Phytophthora* [REICHLE, 1969], and *Nowakowskiella* [CHAMBERS *et al.*, 1967].'

In the few years that have intervened since then, new pictorial examples which tend to support the foregoing point of view have been published. At this time, we prefer to delay providing either an exhaustive list or an analysis of all these new reports on zoospore fine structure. However, since a few of these newly described organelles have been compared specifically with the gamma particles of *B. emersonii*, we would like to give a brief recount of them here.

1. 'Gamma-Like Particles' in the Zoospores of Blastocladiella britannica [CANTINO and TRUESDELL, 1971b]

In spite of the close relationship between this fungus and *B. emersonii*, the gamma-like particles of *B. britannica* differ from the gamma particles of *B. emersonii*; they never yield horseshoe-like profiles when sectioned, and they are recognizably distinct in structure. Whether or not they play a role in zoospore encystment, however, has not been fully determined.

2. 'Inclusion Bodies' in the Anterior Region of the Motile Cells of Coelomomyces punctatus [MARTIN, 1971]

This organism is also a member of the *Blastocladiales*. The 'side body' complex and other aspects of the internal organization in its motile cells greatly resemble structures in the zoospores of *B. emersonii*; so much so, in fact, that MARTIN [1971], suggests that these similarities point to a strong phylogenetic relationship between the two species. Yet, the 'inclusion bodies' in *C. punctatus* apparently do not contain the three-holed boat-shaped matrix so characteristic of the gamma particles in *B. emersonii*; rather, they more nearly resemble the gamma-like particles in *B. britannica*.[11]

[11] However, see the report by MADELIN and BECKETT [1972] regarding the 150 nm diameter, slightly electron-dense, circular profiles in the spores of *Coelomomyces* which are said not to resemble either the gamma particles of *B. emersonii* or the gamma-like particles of *B. britannica*.

3. 'Dense-Body Vesicles' in the Zoospores of
Saprolegnia ferax *and* Dictyuchus sterile [GAY *et al.*, 1971]

It is thought by these investigators that 'dense-body vesicles' may serve as centers of membrane formation during zoospore germination and that, in this sense at least, they resemble the gamma particles of *B. emersonii.* Coincidentally, it is interesting – especially in view of our own earlier thoughts [TRUESDELL and CANTINO, 1970] and most recent position (chapter X, section C, 2) – that GAY *et al.* [1971] believe dense-body vesicles to be sufficiently characteristic as to deserve a status comparable to that of lysosomes and microbodies of higher plants and animals but nonetheless distinct from, and not necessarily homologous with, any of them. Hopefully these dense-body vesicles will be isolated and some of their chemical and enzymatic features established, for the suspected analogy between them and gamma particles could obviously be significantly strengthened – or weakened – thereby.

4. Vesicles in the Motile Cells of Allomyces

When we made our afore-quoted comments about the pictures that had been published of dense particles in planonts of aquatic fungi, we also said: 'Although none of these structures are morphologically identical to gamma particles, variation among the different species of the assemblage of highly diverse creatures known as the water molds would not seem unreasonable. Among them, *Allomyces* is the relative nearest to *B. emersonii*, and its zoospores contain the particles that seem to most nearly resemble gamma particles.' This conclusion was in keeping with the pictures we had seen at that time: electron micrographs of a longitudinal section through a zoospore of *A. macrogynus* [FULLER and CALHOUN, 1968, fig. 1], and several sections through zoospores of *Allomyces javanicus* [MOORE, 1968]; indeed, some of the latter (e.g. in MOORE's fig. 16 and 20) contained profiles of organelles that strikingly resembled the gamma particles we had been studying in *B. emersonii.* But more recently, working with *A. macrogynus* and *A. neo-moniliformis* FULLER and OLSON [1971, p. 181] concluded that 'Gamma bodies, such as those seen in the zoospores of *B. emersonii* [CANTINO and HORENSTEIN, 1956; CANTINO, 1969; CANTINO and MACK, 1969; TRUESDELL and CANTINO, 1970] have not been observed in zoospores of *Allomyces.*' On the other hand, ROBERTSON, [1972, p. 264] says of a 'new' *Allomyces* that '. . . throughout the zoospores are scattered gamma particles similar to those described in *B. emersonii* [REICHLE and FULLER, 1967].' We do not know if he also intended his conclusion to apply to the gamma particles present in our own strain – i.e. the original parent strain – of *B. emersonii*, but from what we can see of the profiles of cytoplasmic particles in ROBERTSON's [1972] figure 5, it is unclear to what degree they resemble the gamma particles depicted in our earliest publication [CANTINO *et al.*, 1963, fig. 16] or in later ones.[12]

It is becoming quite apparent that there is a good deal of interspecific, not to mention intraspecific, variation in the appearance of sections through the electron-dense, unit membrane bound organelles which have thus far been detected in the zoospores of aquatic fungi, and which we have here tried to compare with the gamma particles in *B. emersonii*. When the functions of these organelles, especially those thought to be involved in encystment, have been established using spore suspensions sufficiently synchronized such that zoospore activities can be meaningfully quantified, and when these same organelles have also been isolated and their chemical and enzyme composition partially established, it should become possible to decide whether or not the water molds have distinguished themselves once again by providing the biologist with a new and useful class of subcellular structural elements[13].

[12] Two other reports published after this monograph was written which include information on the appearance and behavior of gamma-like granules should be included here, namely one by A. A. HELD (Canadian J. BOTANY, 51, 1825–1836, 1973) on *Rozella allomycis*, and the other by L. W. OLSON (Protoplasma, 78, 129–144, 1973) on *Allomyces macrogynus*.

[13] It has been proposed at the 2nd. International Fungal Spore Symposium in Provo, Utah, July 1974 (E. C. CANTINO and G. L. MILLS, Form and Function in Chytridiomycete Spores; in Form and Function in the Fungus Spore; ed. D. J. WEBER and W. M. HESS, Wiley Interscience, N. Y.) that the gamma particle be considered as the «type specimen» for a new family of cytoplasmic organelles in the zoospores of aquatic fungi, this new group being named *Encystosomes*. Although it cannot be expected that all encystosomes will be structurally identical, it is expected that they will share certain properties, hence the encystosome has been defined as follows: a cytoplasmic, single membrane bound organelle, containing nucleic acid and chitin synthetase, which is required for chytridiomycete zoospore encystment, one of its direct and essential functions being the production of vesicles which fuse with the plasma membrane during the initial deposition of the cyst wall; in the process, the matrix of the encystosome may be gradually «used up», hence its original identity can eventually be lost.

XIII. References

ALONI, Y. and ATTARDI, G.: Expression of the mitochondrial genome in HeLa cells. II. Evidence for complete transcription of mitochondrial DNA. J. molec. Biol. *55:* 251–270 (1971a).

ALONI, Y. and ATTARDI, G.: Expression of the mitochondrial genome in HeLa cells. IV. Titration of mitochondrial genes for 16S, 12S and 4S RNA. J. molec. Biol. *55:* 271–276 (1971b).

BALDWIN, H. H. and RUSCH, H. P.: The chemistry of differentiation in lower organisms. Annu. Rev. Biochem. *34:* 565–594 (1965).

BARRETT, A. J.: Lysosomal enzymes; in DINGLE Lysosomes, pp. 46–135 (North-Holland, Amsterdam 1972).

BISHOP, D. L. H.; CLAYBROCK, J. R., and SPIEGELMAN, S.: Electrophoretic separation of viral nucleic acids on polyacrylamide gels. J. molec. Biol. *26:* 373–387 (1967).

BLONDEL, B. and TURIAN, G.: Relation between basophilia and fine structure of cytoplasm in the fungus *Allomyces macrogynus Em.* J. biophys. biochem. Cytol. *7:* 127–134 (1960).

BRACKER, C. E.: Ultrastructure of fungi. Annu. Rev. Phytopath. *5:* 343–374 (1967).

BRACKER, C. E.: Cytoplasmic vesicles in germinating spores. Protoplasma, Wien *72:* 381–397 (1971).

BRAWERMANN, G. and EISENSTADT, J. M.: Deoxyribonucleic acid from the chloroplast of *Euglena gracilis.* Biochim. biophys. Acta. *91:* 477–485 (1964).

BRENNER, D. M. and CARROLL, G. C.: Fine-structural correlates of growth in hyphae of *Ascodesmis sphaerospora.* J. Bact. *95:* 658–671 (1968).

BURTON, K.: A study of the condition and mechanism of the diphenylamine reaction for the colorimetric estimation of deoxyribonucleic acid. Biochem. J. *62:* 315–323 (1956).

CABIB, E. and BOWERS, B.: Chitin and yeast budding. J. biol. Chem. *246:* 152–159 (1971).

CABIB, E. and FARKAS, V.: The control of morphogenesis: an enzymatic mechanism for the initiation of septum formation in yeast. Proc. nat. Acad. Sci., Wash. *68:* 2052–2056 (1971).

CABIB, E. and KELLER, F. A.: Chitin and yeast budding. J. biol. Chem. *246:* 167–173 (1971).

CAMARGO, E. P.; DIETRICH, C. P.; SONNEBORN, D., and STROMINGER, J. L.: Biosynthesis of chitin in spores and growing cells of *Blastocladiella emersonii.* J. biol. Chem. *242:* 3121–3128 (1967).

CANTINO, E. C.: Metabolism and morphogenesis in a new *Blastocladiella.* Antonie v. Leeuwenhoek *17:* 325–362 (1951).

CANTINO, E. C.: Intracellular distribution of ^{14}C during sporogenesis in *Blastocladiella emersonii.* Effect of light on hemoprotein. Arch. Mikrobiol. *51:* 42–59 (1965).

CANTINO, E. C.: Morphogenesis in aquatic fungi; in AINSWORTH and SUSSMAN The fungi, vol. 2, pp. 283–337 (Academic Press, New York 1966).

CANTINO, E. C.: The γ-particle, satellite ribosome package, and spheroidal mitochondrion in the zoospore of *Blastocladiella emersonii*. Phytopathology *59:* 1071–1076 (1969).

CANTINO, E. C. and GOLDSTEIN, A.: Bicarbonate-induced synthesis of polysaccharide during morphogenesis by synchronous, single generations of *Blastocladiella emersonii*. Arch. Mikrobiol. *39:* 43–52 (1961).

CANTINO, E. C. and HORENSTEIN, E. A.: Cytoplasmic exchange without gametic copulation in the water mold *Blastocladiella emersonii*. Amer. Naturalist *88:* 143–154 (1954).

CANTINO, E. C. and HORENSTEIN, E. A.: Gamma and the cytoplasmic control of differentiation in *Blastocladiella*. Mycologia *48:* 443–446 (1956).

CANTINO, E. C. and HYATT, M. T.: Phenotypic 'sex' determination in the life history of a new species of *Blastocladiella, B. emersonii*. Antonie v. Leeuwenhoek *19:* 25–70 (1953 a).

CANTINO, E. C. and HYATT, M. T.: Carotenoids and oxidative enzymes in the aquatic Phycomycetes *Blastocladiella* and *Rhizophlyctis*. Amer. J. Bot. *40:* 688–694 (1953b).

CANTINO, E. C. and LOVETT, J. S.: Non-filamentous aquatic fungi: model systems for biochemical studies of morphological differentiation. Adv. Morphogenes. *3:* 33–93 (1964).

CANTINO, E. C.; LOVETT, J. S.; LEAK, L. V., and LYTHGOE, J.: The single mitochondrion, fine structure, and germination of the spore of *Blastocladiella emersonii*. J. gen. Microbiol. *31:* 393–404 (1963).

CANTINO, E. C. and MACK, J. P.: Form and function in the zoospore of *Blastocladiella*. I. The gamma particle and satellite ribosome package. Nova Hedwigia *18:* 115–148 (1969).

CANTINO, E. C. and MYERS, R. B.: Concurrent effect of visible light on gamma particles, chitin synthetase, and encystment capacity in the zoospores of *Blastocladiella emersonii*. Arch. Mikrobiol. *83:* 203–215 (1972).

CANTINO, E. C. and MYERS, R. B.: The γ-particle and intracellular interactions in *Blastocladiella emersonii*. Brookhaven Symp. Biology, No. 25, pp. 51–74 (1974).

CANTINO, E. C. and TRUESDELL, L. C.: Organization and fine structure of the side body and its lipid sac in the zoospore of *Blastocladiella emersonii*. Mycologia *62:* 548–567 (1970).

CANTINO, E. C. and TRUESDELL, L. C.: Cytoplasmic gamma-like particles and other ultrastructural aspects of zoospores of *Blastocladiella britannica*. Trans. brit. mycol. Soc. *56:* 169–179 (1971a).

CANTINO, E. C. and TRUESDELL, L. C.: Light induced encystment of *Blastocladiella emersonii* zoospores. J. gen. Microbiol. *69:* 199–204 (1971b).

CANTINO, E. C.; TRUESDELL, L. C., and SHAW, D. S.: Life history of the motile spores of *Blastocladiella emersonii:* a study in cell differentiation. J. Elisha Mitchell Sci. Soc. *84:* 125–146 (1968).

CHAMBERS, T. C.; MARKUS, K. and WILLOUGHBY, L. G.: The fine structure of the mature zoosporangium of *Nowakowskiella profusa*. J. gen. Microbiol. *46:* 135–141 (1967).

CHAMBERS, T. C. and WILLOUGHBY, L. G.: The fine structure of *Rhizophlyctis rosea*, a soil Phycomycete. J. roy. micr. Soc. *83:* 355–364 (1964).

CHI, J. C. H. and SUYAMA, Y.: Comparative studies on mitochondrial and cytoplasmic ribosomes of *Tetrahymena pyriformis*. J. molec. Biol. *53:* 531–556 (1970).

CH'IH, J. J. and KALF, G. F.: Studies on the biosynthesis of the DNA polymerase of rat liver mitochondria. Arch. Biochem. Biophys. *133:* 38–45 (1969).

COLE, P. E.; YANG, S. K., and CROTHERS, D. M.: Conformational changes of transfer ribonucleic acid. Equilibrium phase diagrams. Biochemistry *11:* 4358–4368 (1972).

COMB, D. G.; BROWN, R., and KATZ S.: The nuclear DNA and RNA components of the aquatic fungus *Blastocladiella emersonii.* J. molec. Biol. *8:* 781–789 (1964).

COMB, D. G. and ZEHAVI-WILLNER, T.: Isolation, purification and properties of 5s ribosomal RNA: a new species of cellular RNA. J. molec. Biol. *23:* 441–458 (1967).

COMB, D. G. and KATZ, S.: Studies on the biosynthesis and methylation af transfer RNA. J. molec. Biol. *8:* 790–800 (1964).

COUDRAY, Y.; QUETIER, F., and GUILLE, E.: New compilation of satellite DNA's. Biochim. biophys. Acta *217:* 259–267 (1970).

DAWID, I. B.: The nature of mitochondrial RNA in oocytes of *Xenopus laevis* and its relation to mitochondrial DNA. Symp. Soc. exp. Biol. *24:* 227–246 (1970).

DAWID, I. B.: Mitochondrial RNA in *Xenopus laevis.* I. The expression of the mitochondrial genome. J. molec. Biol. *63:* 201–216 (1972).

DAWID, I. B. and CHASE, J. W.: Mitochondrial RNA in *Xenopus laevis.* II. Molecular weights and other physical properties of mitochondrial ribosomal and 4S RNA. J. molec. Biol. *63:* 217–231 (1972).

DRLICA, K. A. and KNIGHT, C. A.: Inhibition of chloroplast DNA synthesis by cycloheximide. J. molec. Biol. *61:* 629–641 (1971).

DUBIN, D. T. and BROWN, R. E.: A novel ribosomal RNA in hamster cell mitochondria. Biochim. biophys. Acta *145:* 538–540 (1967).

DUBIN, D. T. and MONTENECOURT, B. S.: Mitochondrial RNA from cultured animal cells. Distinctive high-molecular-weight and 4S species. J. molec. Biol. *48:* 279–295 (1970).

DUTT, M. K.: Changes in pH of Feulgen stain and their effect on the staining of biological material. J. Histochem. Cytochem. *11:* 390–394 (1963).

DUVE, C. DE: Evolution of the peroxisome. Ann. N. Y. Acad. Sci. *168:* 369–381 (1969).

DUVE, C. DE and BERTHET, J.: The use of differential centrifugation in the study of tissue enzymes. Int. Rev. Cytol. *3:* 225–275 (1954).

DUVE, C. DE; PRESSMAN, B. C.; GIANETTO, R.; WATTIAUX, R., and APPLEMANS, F.: Tissue fractionation studies. VI. Biochem. J. *60:* 604–617 (1955).

DUVE, C. DE and WATTIAUX, R.: Functions of lysosomes. Annu. Rev. Physiol. *28:* 435–492 (1966).

EVANS, E. E. and SUSKIND, D.: Characterization of the mitochondrial DNA of the slime mold *Physarum polycephalum.* Biochim. biophys. Acta. *228:* 350–364 (1971).

FAWCETT, D. W.: The cell: its organelles and inclusions (Saunders, Philadelphia 1966).

FLAMM, W. G.; BIRNSTIEL, M. L., and WALKER, P. M. B.: Preparation and fractionation, and isolation of single strands, of DNA by isopycnic ultracentrifugation in fixed-angle rotors; in BIRNIE and FOX Subcellular components, pp. 125–155 (Plenum, New York 1969).

FREDERICK, S. E. and NEWCOMB, E. H.: Cytochemical localization of catalase in leaf microbodies (peroxisomes). J. Cell Biol. *43:* 343–353 (1969).

FREDERICK, S. E.; NEWCOMB, E. H., and VIGIL, E. L.: Fine-structural characterization of plant microbodies. Planta *81:* 229–252 (1968).

FULLER, M. S. and CALHOUN, S. A.: Microtubule-kinetosome relationships in the motile cells of the *Blastocladiales.* Z. Zellforsch. *87:* 526–533 (1968).

Fuller, M. S. and Olsen, L. W.: The zoospore of *Allomyces*. J. gen Microbiol. *66:* 171–183 (1971).

Fuller, M. S. and Reichle, R.: The zoospore and early development of *Rhizidiomyces apophysatus*. Mycologia *57:* 946–961 (1965).

Fuller, M. S. and Reichle, R.: The fine structure of *Monoblepharella sp.* zoospores. Canad. J. Bot. *46:* 279–283 (1968).

Gay, J. L.; Greenwood, A. D., and Heath, I. B.: The formation and behavior of vacuoles (vesicles) during oosphere development and zoospore germination in *Saprolegnia*. J. gen. Microbiol. *65:* 233–241 (1971).

Girbardt, M.: Die Ultrastruktur der Apikalregion von Pilzhyphen. Protoplasma, Wien *67:* 413–441 (1969).

Glaser, L. and Brown, D. H.: The synthesis of chitin in cell-free extracts of *Neurospora crassa*. J. biol. Chem. *228:* 729–742 (1957).

Graves, L. B., jr.; Trelease, R. N.; Grill, A., and Becker W. M.: Localization of glyoxylate cycle enzymes in glyoxysomes in *Euglena*. J. Protozool. *19:* 527–532 (1972).

Grossman, L.; Goldring, E. S., and Marmur, J.: Preferential synthesis of yeast mitochondrial DNA in the absence of protein synthesis. J. molec. Biol. *46:* 367–376 (1969).

Grove, S. N. and Bracker, C. E.: Protoplasmic organization of hyphal tips among fungi: vesicles and spitzenkörper. J. Bact. *104:* 989–1009 (1970).

Grove, S. N.; Bracker, C. E., and Morré, D. J.: An ultrastructural basis for hyphal tip growth in *Pythium ultimum*. Amer. J. Bot. *57:* 245–266 (1970).

Heath, I. B.; Gay, J. L., and Greenwood, A. D.: Cell wall formation in the *Saprolegniales:* cytoplasmic vesicles underlying developing walls. J. gen. Microbiol. *65:* 225–232 (1971).

Held, A. A.: Fungal zoospores are induced to encyst by treatments known to degrade cytoplasmic microtubules. Arch. Mikrobiol. *85:* 209–224 (1972).

Hennessy, S. W. and Cantino, E. C.: Lag-phase sporogenesis in *Blastocladiella emersonii:* induced formation of unispored plantlets. Mycologia *64:* 1066–1087 (1972).

Horenstein, E. A. and Cantino, E. C.: Fungi; in Metz and Monroy Fertilization, vol. 2, pp. 95–133 (Academic Press, New York 1969).

Horgen, P. A.: *In vitro* ribonucleic acid synthesis in the zoospores of the aquatic fungus *Blastocladiella emersonii*. J. Bact. *106:* 281–282 (1971).

Horgen, P. A. and Griffin, D. H.: Specific inhibitors of the three RNA polymerases from the aquatic fungus *Blastocladiella emersonii*. Proc. nat. Acad. Sci., Wash. *68:* 338–341 (1971a).

Horgen, P. A. and Griffin, D. H.: RNA polymerase III of *Blastocladiella emersonii* is mitochondrial. Nature new Biol. *234:* 17–18 (1971b).

Hori, M.; Kakiki, K.; Suzuki, S., and Misato, T.: Studies on the mode of action of polyoxins. III. Relation of polyoxin structure to chitin synthetase inhibition. Agric. biol. Chem. *35:* 1280–1291 (1971).

Horwitz, H. and Holt, C. E.: Specific inhibition by ethidium bromide of mitochondrial DNA synthesis in *Physarum polycephalum*. J. Cell Biol. *49:* 546–553 (1971).

Hruban, Z. and Rechcigl, M.: Microbodies and related particles (Academic Press, New York 1969).

Huang, A. H. C. and Beevers, H.: Isolation of microbodies from plant tissues. Plant Physiol. *48:* 637–641 (1971).

JENSEN, W. A.: Botanical Histochemistry (Freeman, San Francisco 1962).

KATZ, S. and COMB, D. G.: A new method for the determination of the base composition of ribonucleic acid. J. biol. Chem. *238:* 3065–3067 (1963).

KIND, P. R. N. and KING, E. J.: Estimation of plasma phosphatase by determination of hydrolyzed phenol with aminoantipyrine. J. clin. Path. *7:* 322 (1954).

KNITTEL, M. D.; BLACK, C. H.; SANDINE, W. E., and FRASER, D. K.: Use of normal probability paper in determining thermal melting values of deoxyribonucleic acid. Canad. J. Microbiol. *14:* 239–245 (1968).

LAURENT, M. and STEINERT, M.: Electron microscopy of kinetoplastic DNA from *Trypanosoma mega*. Proc. nat. Acad. Sci., Wash. *66:* 419–424 (1970).

LEE, L. F. and BOEZI, J. A.: Characterization of bacteriophage gh-l for *Pseudomonas putrida*. J. Bact. *92:* 1821–1827 (1966).

LENNIE, R. W.; GREGORY, D. W., and BIRT, L. M.: Changes in the nucleic acid content and structure of thoracic mitochondria during development of the blowfly, *Lucilia cuprina*. J. Insect Physiol. *13:* 1745–1756 (1967).

LESSIE, P. E. and LOVETT, J. S.: Ultrastructural changes during sporangium formation and zoospore differentiation in *Blastocladiella emersonii*. Amer. J. Bot. *55:* 220–236 (1968).

LOENING, U. E.: RNA structure and metabolism. Annu. Rev. Plant Physiol. *19:* 37–70 (1968).

LOVETT, J. S.: Chemical and physical characterization of 'nuclear caps' isolated from *Blastocladiella* zoospores. J. Bact. *85:* 1235–1246 (1963).

LOVETT J. S.: Aquatic fungi; in WILT and WESSELLS Methods in developmental biology, pp. 341–358 (Crowell-Collier, New York 1967).

LOVETT, J. S.: Reactivation of ribonucleic acid and protein synthesis during germination of *Blastocladiella* zoospores and the role of the ribosomal nuclear cap. J. Bact. *96:* 962–969 (1968).

LOVETT, J. S. and CANTINO, E. C.: The relation between biochemical and morphological differentiation in *Blastocladiella emersonii*. I. Enzymatic synthesis of glucosamine-6-phosphate. Amer. J. Bot. *47:* 499–505 (1960a).

LOVETT, J. S. and CANTINO, E. C.: The relation between biochemical and morphological differentiation in *Blastocladiella emersonii*. II. Nitrogen metabolism in synchronous cultures. Amer. J. Bot. *47:* 550–560 (1960b).

LOWRY, O. H.; ROSEBROUGH, N. J.; FARR, A. L., and RANDALL, R. J.: Protein measurement with the Folin phenol reagent. J. biol. Chem. *193:* 265–275 (1951).

LUCK, D. J. L. and REICH, E.: DNA in mitochondria of *Neurospora crassa*. Proc. nat. Acad. Sci., Wash. *52:* 931–938 (1964).

LÜCK, H.: Catalase; in BERGMAM Methods of enzymatic analysis, pp. 686–688 (Verlag Chemie, Weinheim 1963).

MADELIN, M. F. and BECKETT, A.: The production of planonts by thin-walled sporangia of the fungus *Coelomomyces indicus*, a parasite of mosquitoes. J. gen. Microbiol. *72:* 185–200 (1972).

MANDEL, M.; IGAMBI, L.; BERGENDAHL, J.; DODSON, M. L., jr., and SCHELTGEN, E.: Correlation of melting temperature and cesium chloride buoyant density of bacterial deoxyribonucleic acid. J. Bact. *101:* 333–338 (1970).

MARCHANT, R. and MOORE, R. T.: Lomasomes and plasmalemmasomes in fungi. Protoplasma *76:* 235–247 (1973).

MARMUR, J.: A procedure for the isolation of deoxyribonucleic acid from microorganisms. J. molec. Biol. *3:* 208–218 (1961).

MARMUR, J. and DOTY, P.: Determination of the base composition of deoxyribonucleic acid from its thermal denaturation temperature. J. molec. Biol. *5:* 109–118 (1962).

MARTIN, W. W.: The ultrastructure of *Coelomomyces punctatus* zoospores. J. Elisha Mitchell sci. Soc. *87:* 209–221 (1971).

MATSUMAE, A. and CANTINO, E. C.: Mitomycin-induced variants of *Blastocladiella emersonii*: alterations in γ-particle content, flagellation, and pigmentation. J. Antibiot. *23:* 162–165 (1970).

MATSUMAE, A.; MYERS, R. B., and CANTINO, E. C.: Comparative numbers of γ-particles in the flagellate cells of various species and mutants of *Blastocladiella*. J. gen. appl. Microbiol. *16:* 443–453 (1970).

MCCLURE, W. K.; PARK, D., and ROBINSON, P. M.: Apical organization in the somatic hyphae of fungi. J. gen. Microbiol. *50:* 177–182 (1968).

MCCURDY, H. D. and CANTINO, E. C.: Isocitritase, glycine-alanine transaminase, and development in *Blastocladiella emersonii*. Plant Physiol. *35:* 463–476 (1960).

MCMANUS, J. F. A. and MOWRY, R. W.: Staining methods (Harper & Row, New York 1960).

MCMURROUGH, I. and BARTNICKI-GARCIA, S.: Properties of a particulate chitin synthetase from *Mucor rouxii*. J. biol. Chem. *246:* 4008–4016 (1971).

MCMURROUGH, I.; FLORES-CARREON, A., and BARTNICKI-GARCIA, S.: Pathway of chitin synthesis and cellular localization of chitin synthetase in *Mucor rouxii*. J. biol. Chem *246:* 3999–4007 (1971).

MERRICK, W. C. and DURE, L. S., III: The developmental biochemistry of cotton seed embryogenesis and germination. J. biol. Chem. *247:* 7988–7999 (1972).

MIMS, C. W.: Spore-wall formation in the myxomycete *Arcyria cinerea*. Trans brit. mycol. Soc. *59:* 477–481 (1972).

MIURA, K.: Preparation of bacterial DNA by the phenol-pH 9-RNAse method; in GROSSMAN and MOLDOVE Methods in enzymology, vol. 22/A, pp. 543–544 (Academic Press, New York 1967).

MOLLENHAUER, H. H.; MORRÉ, D. J., and KELLY, A. G.: The widespread occurrence of plant cytosomes resembling animal microbodies. Protoplasma *62:* 44–52 (1966).

MOORE, R. T.: Fine structure of mycota. XIII. Zoospore and nuclear cap formation in *Allomyces*. J. Elisha Mitchell sci. Soc. *84:* 147–165 (1968).

MÜLLER, M.: Peroxisomes of protozoa. Ann. N. Y. Acad. Sci. *168:* 292–301 (1969).

MYERS, R. B. and CANTINO, E. C.: DNA profile of the spore of *Blastocladiella emersonii:* evidence for γ-particle DNA. Arch. Mikrobiol. *78:* 252–267 (1971).

NASS, M. M. K.: Mitochondrial DNA: advances, problems and goals. Science *165:* 25–35 (1969a).

NASS, M. M. K.: Mitochondrial DNA. II. Structure and physicochemical properties of isolated DNA. J. molec. Biol. *42:* 529–545 (1969b).

NASS, M. M. K. and BUCK, C. A.: Studies on mitochondrial tRNA from animal cells. II. Hybridization of aminoacyl-tRNA from rat liver mitochondria with heavy and light complementary strands of mitochondrial DNA. J. molec. Biol. *54:* 187–198 (1970).

NORRMAN, J.; MYERS, R. B.; GIDDINGS, T. H., and CANTINO, E. C.: Partial purification of L-glutamine: D-fructose 6-phosphate aminotransferase from zoospores of *Blastocladiella emersonii*. Biophys. biochim. Acta *302:* 173–177 (1973).

PESTKA, S.: Inhibitors of ribosome functions. Annu. Rev. Microbiol. *25:* 487–562 (1971).

PORTER, C. A. and JAWORSKI, E. G.: The synthesis of chitin by particulate preparations of *Allomyces macrogynus*. Biochemistry *5:* 1149–1154 (1966).

QUAIL, P. H. and SCANDALIOS, J. G.: Turnover of genetically defined catalase isozymes in maize. Proc. nat. Acad. Sci., Wash. *68:* 1402–1406 (1971).

REICHLE, R. E.: Fine structure of *Phytophthora parasitica*. Mycologia *61:* 30–51 (1969).

REICHLE, R. E. and FULLER, M. S.: The fine structure of *Blastocladiella emersonii* zoospores. Amer. J. Bot. *54:* 81–92 (1967).

ROBERTSON, J. A.: Phototaxis in a new *Allomyces*. Arch. Mikrobiol. *85:* 259–266 (1972).

SAGER, R.: Cytoplasmic genes and organelles (Academic Press, New York 1972).

SCHIFF, J. A. and EPSTEIN, H. T.: The continuity of the chloroplasts in *Euglena*; in LOCKE Reproduction: molecular, subcellular and cellular. 24th Annu. Symp. Soc. Development Growth (Academic Press, New York 1965).

SCHILDKRAUT, C. L.; MARMUR, J., and DOTY, P.: Determination of the base composition of deoxyribonucleic acid from its buoyant density in CsCl. J. molec. Biol *4:* 430–443 (1962).

SCHMID, P.; SCHMID, C., and BRODIE, D. C.: The determination of the total deoxyribose of deoxyribonucleic acid. J. biol. Chem. *238:* 1068–1072 (1963).

SCHMOYER, I. R. and LOVETT, J. S.: Regulation of protein synthesis in zoospores of *Blastocladiella*. J. Bact. *100:* 854–864 (1969).

SCHNEIDER, W. C.: Determination of nucleic acid in tissue by pentose analysis; in COLOWICK and KAPLAN Methods in enzymology, vol. 3, pp. 680–682 (Academic Press, New York 1957).

SCHWEIGER, H. G.: Synthesis of RNA in *Acetabularia*. Symp. Soc. exp. Biol. *24:* 327–344 (1970).

SHAW, D. S. and CANTINO, E. C.: An albino mutant of *Blastocladiella emersonii*: comparative studies of zoospore behavior and fine structure. J. gen. Microbiol. *59:* 369–382 (1969).

SISLER, H. D. and SIEGEL M. R.: Cycloheximide and other glutarimide antibiotics; in GOTTLIEB and SHAW Antibiotics: mechanism of action, vol. 1, pp. 283–307 (Springer, New York 1967).

SMITH, J. E. and GALBRAITH, J. C.: Biochemical and physiological aspects of differentiation in the fungi. Adv. microbial Physiol. *5:* 45–134 (1971).

SOLL, D. R.; BROMBERG, R., and SONNEBORN, D. R.: Zoospore germination in the water mold, *Blastocladiella emersonii*. I. Measurement of germination and sequence of subcellular morphological changes. Develop. Biol. *20:* 183–217 (1969).

SOLL, D. R. and SONNEBORN, D. R.: Zoospore germination in the water mold, *Blastocladiella emersonii*. II. Influence of cellular and environmental variables on germination. Develop. Biol. *20:* 218–235 (1969).

SOLL, D. R. and SONNEBORN, D. R.: Zoospore germination in *Blastocladiella emersonii*: cell differentiation without protein synthesis. Proc. nat. Acad. Sci., Wash *68:* 459–463 (1971a).

SOLL, D. R. and SONNEBORN, D. R.: Zoospore germination in *Blastocladiella emersonii*.

III. Structural changes in relation to protein and RNA synthesis. J. Cell Sci. *9:* 679–699 (1971b).

SOLL, D. R. and SONNEBORN, D. R.: Zoospore germination in *Blastocladiella emersonii.* IV. Ion control over cell differentiation. J. Cell Sci. *10:* 315–333 (1972).

STUTZ, E. and RAWSON, J. R.: Separation and characterization of *Euglena gracilis* chloroplast single-strand DNA. Biochim. biophys. Acta *209:* 16–23 (1970).

SUBERKROPP, K. F. and CANTINO, E. C.: Environmental control of motility and encystment in *Blastocladiella emersonii* zoospores at high population densities. Trans. brit. mycol. Soc. *59:* 1–13 (1972).

SUBERKROPP, K. F. and CANTINO, E. C.: Developmental physiology of the zoospores of *Blastocladiella emersonii.* Arch. Mikrobiol. *89:* 205–221 (1973).

SUYAMA, Y. and MIURA, K.: Size and structural variation of mitochondrial DNA. Proc. nat. Acad. Sci., Wash. *60:* 235–242 (1968).

SWIFT, H.: Cytochemical techniques for nucleic acids; in CHARGAFF and DAVIDSON The nucleic acids, vol. 2, pp. 51–92 (Academic Press, New York 1955).

SWIFT, H.: in ALLEN The molecular control of cellular activity, p. 73 (McGraw-Hill, New York 1962).

SWIFT, H. and WOLSTENHOLME, D.R.: Mitochondria and chloroplasts: nucleic acid and the problem of biogenesis; LIMA DE FARIA Handbook of molecular cytology, pp. 972–1146 (North-Holland, Amsterdam 1969).

TANGUAY, R. and CHAUDHARY, K. D.: Studies on mitochondria. I. Mitochondrial DNA in relation to morphological development of sarcosomes in thoracic muscles of the African desert locust (*Schistocerca gregaria Forsk*). Canad. J. Biochem. *49:* 357–367 (1971).

TEWARI, K. K.: Genetic autonomy of extra-nuclear organelles. Annu. Rev. Plant Physiol. *22:* 141–168 (1971).

TOLBERT, N. E. and YAMAZAKI, R. K.: Leaf peroxisomes and their relation to photorespiration and photosynthesis. Ann. N. Y. Acad. Sci. *168:* 325–341 (1969).

TRUESDELL, L. C. and CANTINO, E. C.: Decay of γ-particles in germinating spores of *Blastocladiella emersonii.* Arch. Mikrobiol. *70:* 378–392 (1970).

TRUESDELL, L. C. and CANTINO, E. C.: The induction and early events of germination in the zoospore of *Blastocladiella emersonii.* Curr. Topics develop. Biol. *6:* 1–44 (1971).

TRUESDELL, L. C. and CANTINO, E. C.: Myelin-like 'artifacts' in the zoospores of *Blastocladiella emersonii.* Trans. brit. mycol. Soc. *59:* 129–132 (1972).

TURIAN, G.: Différenciation fongique. Monogr. Physiologie Végétale, no. 5 (Masson, Paris 1969).

VAN BRUGGEN, E. F. J.; RUNNER, C. M.; BORST P.; RUTTENBERG, G. J. C. M.; KROON, A. M., and SCHUURMANS STEKHOVEN, F. M. A. H.: Mitochondrial DNA. III. Electron microscopy of DNA released from mitochondria by osmotic shock. Biochim. biophys. Acta *161:* 402–414 (1968).

VIGIL, E. L.: Cytochemical and developmental changes in microbodies (glyoxysomes) and related organelles of castor bean endosperm. J. Cell Biol. *46:* 435–454 (1970).

VINOGRAD, J. and HEARST, J. E.: Equilibrium sedimentation of macromolecules and viruses in a density gradient; in ZECHMEISTER Fortschr. Chem. org. Naturstoffe, vol. 20, p. 372 (Springer, Vienna 1962).

Werry, P. A. T. J. and Wanka, F.: The effect of cycloheximide on the synthesis of major and satellite DNA components in *Physarum polycephalum*. Biochim. biophys. Acta *287:* 232–235 (1972).

Wolstenholme, D. R. and Gross, N. J.: The form and size of mitochondrial DNA of the red bean, *Phaseolus vulgaris*. Proc. nat. Acad. Sci., Wash. *61:* 245–252 (1968).

Woodcock, C. L. F. and Bogorod, L.: Evidence for variation in the quantity of DNA among plastids of *Acetabularia*. J. Cell Biol. *44:* 361–375 (1970).

Wyatt, G. R.: Recognition and estimation of 5-methyl cytosine in nucleic acids. Biochem. J. *48:* 581–584 (1951).

Zehavi-Willner, T. and Danon, D.: The isolation and properties of reticulocyte 5S RNA. FEBS Letters *26:* 151–156 (1972).